Environmental Postcolonialism

Ecocritical Theory and Practice
Series Editor: Douglas A. Vakoch, METI

Advisory Board
Sinan Akilli, Cappadocia University, Turkey; Bruce Allen, Seisen University, Japan; Zélia Bora, Federal University of Paraíba, Brazil; Izabel Brandão, Federal University of Alagoas, Brazil; Byron Caminero-Santangelo, University of Kansas, USA; Chia-ju Chang, Brooklyn College, The City College of New York, USA; H. Louise Davis, Miami University, USA; Simão Farias Almeida, Federal University of Roraima, Brazil; George Handley, Brigham Young University, USA; Steven Hartman, Mälardalen University, Sweden; Isabel Hoving, Leiden University, The Netherlands; Idom Thomas Inyabri, University of Calabar, Nigeria; Serenella Iovino, University of Turin, Italy; Daniela Kato, Kyoto Institute of Technology, Japan; Petr Kopecký, University of Ostrava, Czech Republic; Julia Kuznetski, Tallinn University, Estonia; Bei Liu, Shandong Normal University, People's Republic of China; Serpil Oppermann, Cappadocia University, Turkey; John Ryan, University of New England, Australia; Christian Schmitt-Kilb, University of Rostock, Germany; Joshua Schuster, Western University, Canada; Heike Schwarz, University of Augsburg, Germany; Murali Sivaramakrishnan, Pondicherry University, India; Scott Slovic, University of Idaho, USA; Heather Sullivan, Trinity University, USA; David Taylor, Stony Brook University, USA; J. Etienne Terblanche, North-West University, South Africa; Cheng Xiangzhan, Shandong University, China; Hubert Zapf, University of Augsburg, Germany

Ecocritical Theory and Practice highlights innovative scholarship at the interface of literary/cultural studies and the environment, seeking to foster an ongoing dialogue between academics and environmental activists.

Recent Titles
Environmental Postcolonialism: A Literary Response edited by Shubhanku Kochar and M. Anjum Khan
Reading Aridity in Western American Literature edited by Jada Ach and Gary Reger
Reading Cats and Dogs: Companion Animals in World Literature edited by Françoise Besson, Zelia M. Bora, Marianne Marroum, and Scott Slovic
Turkish Ecocriticism: From Neolithic to Contemporary Timescapes edited by Sinan Akilli and Serpil Oppermann
Avenging Nature: The Role of Nature in Modern and Contemporary Art and Literature, edited by Eduardo Valls Oyarzun, Rebeca Gualberto Valverde, Noelia Malla Garcia, María Colom Jiménez, and Rebeca Cordero Sánchez
Migrant Ecologies: Zheng Xiaoqiong's Women Migrant Workers by Zhou Xiaojing
Climate Consciousness and Environmental Activism in Composition: Writing to Save the World edited by Joseph R. Lease
Rethinking Nathaniel Hawthorne and Nature: Ecocriticism and the Tangled Landscape of American Romance by Steven Petersheim
The Poetics and Politics of Gardening in Hard Times edited by Naomi Milthorpe
The Human–Animal Boundary: Exploring the Line in Philosophy and Fiction edited by Mario Wenning and Nandita Batra
Dwellings of Enchantment: Writing and Reenchanting the Earth edited by Bénédicte Meillon

Environmental Postcolonialism

A Literary Response

Edited by
Shubhanku Kochar and M. Anjum Khan

LEXINGTON BOOKS
Lanham • Boulder • New York • London

Published by Lexington Books
An imprint of The Rowman and Littlefield Publishing Group, Inc.
4501 Forbes Boulevard, Suite 200, Lanham, Maryland 20706
www.rowman.com

6 Tinworth Street, London SE11 5AL, United Kingdom

Copyright © 2021 by The Rowman and Littlefield Publishing Group, Inc.

All rights reserved. No part of this book may be reproduced in any form or by any electronic or mechanical means, including information storage and retrieval systems, without written permission from the publisher, except by a reviewer who may quote passages in a review.

British Library Cataloguing in Publication Information Available

Library of Congress Control Number: 2020949032

ISBN 9781793634566 (cloth) | ISBN 9781793634573 (epub)
ISBN 9781793634580 (pbk)

Contents

1 Introduction — 1
 Shubhanku Kochar and M. Anjum Khan

2 "Transformation Is the Rule of Life": Environment and the Search for Utopia in *The Hungry Tide* — 17
 Suzy Woltmann

3 Cultural Nationalism and Sacred Groves of Kerala — 31
 Anupama Nayar

4 Politics, Oil, and Theater in Africa — 47
 Stephen Ogheneruro Okpadah

5 Through the Postcolonial Lens: Reading the Environment in Narratives from India's North East — 61
 Kalpana Bora Barman

6 "Aesthetics of Belonging": Construction of a Postcolonial Landscape in Daud Kamal's Poetry — 75
 Humaira Riaz

7 *I Am a Tree Leaning*: Neocolonialism, Eco-consciousness, and the Decolonized Self in Margaret Atwood's *Surfacing* — 89
 Anik Sarkar

8 For Appearances Must Deceive: Misreading the Environment in *Days and Nights in the Forest* and Its Cinematic Adaptation — 103
 Chinmaya Lal Thakur

Contents

9   Postcolonial Ecology and Representation: Exploring "Ashani Sanket" as an Ecofilm   115
Neepa Sarkar

10   Land, Labor, and Family: The Impact of US Colonization on Puerto Rico in Esmeralda Santiago's *When I Was Puerto Rican*   127
Renée Latchman

11   "Coloniality" of Humans and the Ecology: An Ecocritical Reading of Shubhangi Swarup's *Latitudes of Longing*   139
Risha Baruah

12   Women and Power: Digital Cameras in Postcolonial Caribbean Spaces in Literature   155
Denise M. Jarrett

13   Nature and Resistance in Coetzee and Abani: The Transcoporeal in African Fiction   171
Puja Sen Majumdar

14   Colonialism, Capitalism, and Nature: A Study of Alex Haley's *Roots* and Ngũgĩ wa Thiong'o's *Petals of Blood*   183
Shivani Duggal

15   Beyond the Dichotomy of Humans and Animals: Situating Ecology in Coetzee's Writings   197
Bipasha Mandal

16   Provincializing Ecocriticism: Postcolonial Ecocritical Thoughts and Environmental-Historical Difference   211
Animesh Roy

Index   227

About the Contributors   229

Chapter 1

Introduction

Shubhanku Kochar and M. Anjum Khan

SOCIOHISTORICAL AND LITERARY CONTEXT

Human history, as Rodney explains in *How Europe Underdeveloped Africa*, on this Earth commences with movement/migration. Our ancestors were hunters and gatherers initially. They used to roam here and there for food and shelter. Their life was defined by mobility and uncertainty. Wherever they would find games and pastures, they would pitch their tents. Whenever their supplies would run out, they would move again. But, with the passage of time, human beings learned husbandry. As Rodney illustrates, our history is the history of our coming to terms with our environment. Human civilizations have passed from one stage to another with respect to their capacity to interact with their environment.

Once agriculture became a defining factor, our ancestors started to live a stable life. They started to cultivate and felt that there was no need to walk all the time. They began to own families thereby giving birth to a stable society.

For example, Ngugi Wa Thiong'o in *Petals of Blood* narrates the story of how the village Ilmorog in Kenya was once an extreme wilderness and how it became a hub of power and glory. He narrates how once there was no one living in Ilmorog. How there was jungle all around. How there came one strong man with his cattle and how he decides to settle there in Ilmorog. He was a strong and sturdy man. With his sweat and blood, he builds Ilmorog. He felled the trees, cleared the ground, and started tilling with his followers. Soon there was a bumper harvest, and soon his name and fame traveled far and wide. Soon, he found many wives and became the leader and chief of the place.

This process of settling was different from colonialism in manner and method, though it looks similar. Firstly, to bring Rodney again, this was a ubiquitous process. Primarily, almost all societies were communal. Then after hundreds of years, they passed into feudal faze by natural development. After feudalism, capitalism and socialism became way of life depending upon the internal and external factors. Secondly, this was a survival strategy adopted by man. This meant, he was trying to explore the possibility of life. He was not intruding into anyone's territory. He was not stripping land for cupidity. He was not exploiting nature and the other for personal gains, whereas colonialism included all this and many more.

As Fisher argues in *An Environmental History of India*, colonialism was motivated by greed, so the colonizers did not care much for their human and nonhuman subjects. For them, men and land or culture and nature were nothing, but alike, as both were disregarded. From the fifteenth to the second half of the twentieth century, the saga of the destruction of both human and nature remained unparalleled. While considerable scholarship has already been invested upon how colonialism destroyed and exterminated human life and its various institutions, what present work aims to bring to for front is how colonialism was equally detrimental for ecology in the ex-colonies. The present book also highlights how various ecosystems were either pulverized or rendered vulnerable as a result of centuries of colonial occupation.

In 1598, a boat full of whites arrived in what is today known as Mauritius. They were Dutch sailors who were in search of new territories. They decided to camp on the newly discovered land. They had also brought dogs and rats along. Slowly, they began to settle there. They found a beautiful bird Dodo there. The bird was feathery and pretty. It did not exhibit any fear of human beings. As a result, in 100 years, it disappeared. It was eaten up by the white settlers. After 1700 years, it was never seen. Both the colonizers and their dogs hunted and ate it mercilessly. There are many such examples where one discovers how the colonizers changed ecologies and various ecosystems unmindfully for their selfish gains. In fact, it would not be an exaggeration if one argues that many ecological disasters that one encounters today are the product of imperial invasions, capitalist desires, and colonial policies.

A well-known poem, "Who?" by Alice Walker can be cited, to begin with. She deploys irony and sarcasm to indict the Europeans for their duel destruction of both culture and nature. She writes:

Who has not been? Invaded by the Wasichu?
Not I, said the people.
Not I, said the trees. (2010, 55)

As these lines indicate, the colonizers had no soft corner either for the human world or for the natural world. They eliminated both the entities for their own profit. As history informs us when Spaniards arrived in Central and

South America in the early sixteenth century, they had to confront indigenous communities. To establish their kingdom, they carried out the massacre on hitherto unprecedented scale. Countless natives were murdered. Villages after villages were vandalized. It surely destroyed countless human habitats and ecosystems. Natives of Central and South America were addressed as Indians. They were people with animistic beliefs. They used to live in proximity to nature. When their habitat was bulldozed, it also meant that their source of sustenance was also obliterated. Their trees, huts, and animals, along with their weapons, were burned just to establish the dominance of the Europeans. After the natives were either wiped away or were made to leave, the land was reshaped as per the Spanish way of life.

It was not very difficult for the whites to harm nature. Their culture gave them a license for that. As Lynn White Jr. in "The Historical Roots of our Ecological Crisis" opines that Christianity bears a huge burden of guilt. He further states that Christian society is fed upon the teachings of the *Bible*, where God grants absolute power to man, His son, to do anything with his surroundings including the natural world. He is told that nature exists only to cater to his needs. This decree of God has been taken very seriously by the European colonizers. Armed with this belief, as Lynn White continues, white settlers have been decimating sacred groves for past 300 years. For any westerner, a tree is just a tree, a corporeal presence. He can harm it to serve his desires, whereas, for Indians and Africans, the tree is an embodiment of the divine spirit. People living in ex-colonies had been worshiping nature for ages. For example, Buddhism and Jainism in the Indian subcontinent and animism in the African culture made it impossible for the practitioners to hurt nature.

Christopher Manes in "Nature and Silence" points out yet another factor motivating the mindset of the Europeans. He states that in the European scheme of things "the great chain of being" plays a very important role. According to which, man is at the center of the scheme, whereas God is the head followed by various celestial beings in their respective hierarchy. Man forms the center followed by flora and fauna. This unique position gives him authority over nature. He further maintains that the birth of Christian exegeses was a stepping stone, which made it possible for man to impose his own understanding of natural occurrences. It was followed by the arrival of humanism and empirical science in Europe that granted the Europeans undisputed power over the natural environment.

When Europeans arrived in their intended colonies, they were armed with this ideological apparatus. They also had developed the notion of self-esteem because of guns and teachings of the *Bible*. It was not very difficult for them to sabotage culture and nature. Their greed, coupled with their chicanery, made matters verse. To provide themselves with absolute control, they embarked upon the mission of subduing both man and nature. They started molding both human mind and natural landscape as per their need and European standards.

As Shashi Tharoor explains in *Era of Darkness*, Britishers in India did nothing to benefit the Indians. They were representatives of capitalism. Their own mercenary motives were always their prime focus. They would send British young men to India as administrative officers, and they would go back as aristocrats to England after squeezing both human and nonhuman resources. For example, as Tharoor upholds, Indian forests were uprooted heavily by the Britishers for the following factors. First, they needed to lay railway tracks for commercial and administrative needs. Second, to provide the wooden benches and births in trains, they again needed heavy wood of the trees like Teak and Deodar. Third, to make furniture for the English men back home, they extracted a heavy amount of timber. They encouraged the plantation of cash crops like cotton, opium, and tea. To plant tea, they uprooted many jungles and converted the newly vacant place into tea plantations. They also established botanical gardens to encourage the production of drought-resistant plants. Their aim was not to beautify India by establishing botanical gardens, but they were tightening their noose upon Indian natural resources by European technology.

To plant and cultivate opium, they had to uproot many trees and clear jungles. To safeguard their crops, they encouraged hunting of the wild animals on a hitherto unimagined scale. It is not that hunting was not a known thing in the Indian subcontinent. In fact, many kings and princes from ancient and mogul empire were well-known hunters of their times. As Fisher records that Akbar and his successors were fond of hunting. Akbar was an expert in taming wild elephants, whereas his son, Jahangir, was also like his father. Actually, in the past, hunting was associated with power, authority, and masculinity. It was a favorite game of the aristocrats. Many Indian folk tales are inundated with kings often losing their way in the forests while on a hunting spree. Great epics like *Ramayana* and *Mahabharat* are full of the great heroes who are skilled hunters. But the colonizers permitted hunting to serve their personal profit. Tharoor remarks that many Britishers themselves were hunters, and they played their games with Indian princes and landlords. It was a matter of prestige to accompany or host a Britisher in those days. Because, as Fisher reasserts, many princes who were allowed to sit on their throne on the condition that they will let the Britishers do what they wanted, had nothing to do, but hunting.

Because these princes had nothing to do, as they had the services of British Indian army, they ended up becoming effeminate. For them, hunting was the only place to show their valor and masculinity. The result of all this was that by the beginning of the twentieth century, many animals like elephants, lions, and cheetahs were almost extinct. Thanks to the rulers of Junagarh who exercised some prudence and preserved Gir forest that we are left with some lions and cheetahs.

Sri Aurobindo in his famous poem "The Tiger and the Deer" vividly presents this phenomenon. He paints a graphic picture of a peaceful jungle where the ecosystem is working smoothly. The food chain is observed. The deer is eaten by the tigers. The jungle is in order. But he ends the poem with skepticism when he writes:

Brilliant, crouching, slouching, what crept through the green heart of the forest,
Gleaming eyes and mighty chest and soft soundless paws of grandeur and murder?
The wind slipped through the leaves as if afraid lest its voice and
   the noise of its steps perturb the pitiless Splendour,
Hardly daring to breathe. But the great beast crouched and crept,
   and crept and crouched a last time, noiseless, fatal,
Till suddenly death leaped on the beautiful wild deer as it drank
Unsuspecting from the great pool in the forest's coolness and shadow,
And it fell and, torn, died remembering its mate left sole in the deep woodland, --
Destroyed, the mild harmless beauty by the strong cruel beauty in Nature.
But a day may yet come when the tiger crouches and leaps
   no more in the dangerous heart of the forest,
As the mammoth shakes no more the plains of Asia;
Still then shall the beautiful wild deer drink from the
   coolness of great pools in the leaves' shadow.
The mighty perish in their might,
The slain survive the slayer. (Gokak 1970, 129)

All the policies implemented by Britishers were indeed harmful to both Indians and their natural landscape. For example, by the end of the Raj, an almost entire nation was connected by the train, but in order to achieve this, countless jungles were deforested, many kilometers of mountains were blasted, and many rivers were to be subdued. Humayun Kabir in his popular poem "Trains" tries to capture this disaster through the eyes of a child. His intention is to critique the impact of modernization on flora and fauna, but he presents it through the innocence of the child narrator. The protagonist questions her mother:

Where do all these trains go day and night? You say they bore their way through hills,
   They roar over bridges across mighty streams, They crash through forests and vast plains, But at the end of their restless journeyings—Where do they go and finally rest? (Gokak 1970, 232)

Likewise, the task of the forest department during the Raj was to establish the control of the state over forests so that timbering could go unhampered. Many times, the forest officials who were chiefly Britishers, at least before World War I, came into clash with the forest dwellers who were living and sustaining in the jungles for ages, but they were ousted as tribes and savages. Similarly, the aim behind the canal project was not to make India rich but to force the land to produce maximum yield, thereby increasing the profit of the colonizers. The dams were also built to facilitate more and more revenues. During the time period of 1757 and 1947, India had to witness many droughts, floods, and epidemics. As a result, along with the deaths of human beings, countless birds, animals, trees, and other natural elements disappeared, whose count is still pending.

Wherever the Europeans went to colonize, their aim was to churn both human and nonhuman alike for wealth. It would not be wrong if one concludes that the industrial revolution in Europe was sponsored by the resources extracted from Asia and Africa. Their disregard for the land of the natives was such that they mapped it for better control. In Asia and Africa, natives used to respect land to the point of calling her mother. For illustration, one can cite *Weep Not Child* by Ngugi Wa Thiong'o, where the writer shows how Ngotho, a poor laborer, who works on the land that now belongs to the district officer. The land originally belonged to Ngotho and his family. But it has been usurped by the Britishers. Only a few farmers of Kenya were allowed to hold their land that too if they agreed to cultivate cash crops. Ngotho's devotion to work upon his own land is sacrosanct. He worships his land as if she is a deity. The colonizers had no regard for all that. For them, business mattered the most and there was no place for fellow feelings.

Chinua Achebe also in *Things Fall Apart* demonstrates how natives of Igbo community regarded nature as alive. They had the concept of sacred groves. They never transgressed the boundaries of the forest for fear and respect. But, the arrival of the whites changed everything. The colonizers wanted to build their church and school, and they started clearing the land reserved for the forest, which was untouched by the natives for ages. In the same text, Achebe unfolds how the Priests in the church made fun of the Gods of the natives because they used to believe in the spirits of nature.

Alice Walker also in *The Color Purple* reveals how in Africa the peaceful lives of the natives were first intruded and later finished by the white masters. The natives of the village where Nettie goes as part of her missionary work were living in tandem with nature. They derived their substance and sustenance from nature. But this relation with nature is altered by the arrival of the colonizers. A village chief tells his listeners when he comes back from his meeting with the new masters that the land was naked and flat as the head of a bald man. Nettie records in her letters that the natives were forced to plant rubber in place of coconut and cassava that they were cultivating for generations. They are quite unhappy because they are compelled to plant what they cannot eat. They had to give rent to live on their own land. They had to pay money to buy water for their daily needs. Everything that they were using free of cost as nature's bounty had been confiscated by the colonizers; as a result, their relation with their environment had changed completely and they were not happy about it.

Joseph Conrad, likewise, in *Heart of Darkness* exposes the cruelty of the colonizers in Congo. There the officers are appointed to extract ivory from the heart of the forest. More the yield, more are the chances of promotion. There natives are tortured on the gunpoint to collect rubber and ivory, which are exported to European markets. The text is a clear example of how the colonizers worked with the motto of gaining more, and to achieve this, they were ready to exploit both man and nature. In *Heart of Darkness*, both natives and their natural surrounding are presented as pre-historic because Conrad himself was a European and his sympathies were with the whites, but he was courageous enough to expose the institution of colonialism with its bleak and gloomy aspects.

Rodney blasts the myth of good and social work done by the Europeans in Africa in his seminal book *How Europe Underdeveloped Africa*. He concludes that after years of exploitation, Africa had nothing except penury and civil wars. He contends that centuries of oppression left Africa with scarred land and wide holes. Europeans plundered her gold, diamonds, rubber, ivory, palm oil, cocoa, cotton, coffee, and many more things. For centuries, Africans were abducted to work as slaves on American plantations. There, they were forced to produce cash crops like sugarcane and cotton. Rodney affirms that European industry, towns, pavements, skyscrapers, and entire Europe's infrastructure were sponsored by African and Asian resources in such a manner that when the colonizers left their erstwhile colonies, there was a vast chasm betwixt Europe and North America on the one side and Africa and Asia on the other side.

CONCEPTUAL FRAMEWORK

We are all part of a common world, but one that is changing rapidly for the immediate benefit of some at the expense of a great many others. (Feder 2014, 6)

The present volume is very important in these capitalistic times, where the main objective is profit and self-interest. It is as important as to look back, as to look ahead. The so-called developing and underdeveloped countries are still suffering from the brunt of colonialism in terms of depleted and polluted natural resources. Further, this critical intervention is a form of activism propagating environmental justice against the overwhelming capitalism. Capitalism is an extension of colonialism in many ways. Only the colonizer has changed.

The interphase between material and ideas also forms the key ingredient of this inquiry. The transition from an old order to a new order is also a constituent in this regard. The interplay between environment and politics is the key to understand this book. "A postcolonial approach to the environmental humanities is self-reflexive in its engagement with histories and knowledges of ecological difference, particularly at an institutional level, where universities are implicated" (DeLoughrey et al. 2015, 3).

Environment is both material and biological. It is available in the form of resources. Therefore, it becomes a commodity, which is used, spent, and consumed. Further, it is also biological and living. In fact, it is as alive as the humans. Human experience cannot be devoid of nature. "The narrative is less about a personal journey and more of an exposé of ecologies, a treatise of the way that humans are co-inhabitants of the world" (Iheka 2018, 29).

The identity of an individual is closely linked with land, which is a natural resource. The symbiosis between subjectivity and land has been achieved in imperialism. Undeniably, the land has been the chief imperial motive. "The concepts of frontier and wilderness have undergirded the colonial enterprise, and have justified the occupation of 'empty' continents. These concepts, beginning in the early 1400s, were invoked to authorize genocide and slavery in the Americas and Africa on a grand scale" (Monani and Adamson 2017, 3).

The indigenous culture and the indigenous nature go hand-in-hand. It can also be said that both are the same. Human-animal dichotomy or binary is an important aspect of postcolonial ecocriticism. For the colonizer, the line dividing this binary is erased, and both the human native and animal become the same. The binary of body and language is also interrelated. The natural environment and animals' corporeal quality against the colonizers' linguistic power is a point of reflection. The natives who also form the part of the environment have a body but no expression or voice, or at least not important and worth listening. Therefore, they also become mute like the environment they live in. Further the binary of colonizer and colonized, and colonizer and colonies are also essential to understand the structure of exploitation.

Agency is also important to understand postcolonial ecocriticism. As both the colonized and the environment were reduced as same. Both were powerless and became mere resources in the hands of colonizers. "The field of postcolonial studies has long been engaged with questions of agency and representation of the nonspeaking or subaltern subject, foregrounding the ways in which narrative and language effectively displace the production of difference and alterity" (DeLoughrey and Handley 2011, 27).

Bioethics is an emerging argument. However, it has always been there from the time immemorial. The famous historical anecdote of Manunidhi Cholan is an example of the inherent bioethics in the East. The famous king, Manunidhi Cholan, executes his own son for running over a calf in order to render justice to the mother cow. The environment ought not to be marginalized; instead, it ought to be considered as formative and foundational for the human society.

The hegemonic triumph is a condition that defines the colonial regime. The colonizer's dominance is over the colonies—both humans and natural resources. Invasion, encroachment, and taming are some of the colonial acts. Especially, exploitation underlines all of these. This colonial behavior carries on in postcolonial era as well. Therefore, it becomes essential to understand and examine the predicament of the environment in postcolonial times. Postcolonialism is most anthropocentric in nature. Environmental justice is marginalized. The industrial and political ambitions are colonial products. However, these remain to exist in the postcolonial era as well.

The idea of monopolizing is deep-rooted in colonialism. Though the times have changed from colonial to postcolonial, the remains of exploitation and monopolization still brew the present condition. Colonialism has been a political act. It has also been a socioeconomic and political act. Ironically, it began as a commercial operation and it metamorphosed into a political enterprise. African history is an integral part of postcolonial histories. Toyin Falola and Emily Brownell begin their book:

> Environmental history and African history are products of the same historical moment of the 1960s. Yet, whereas environmental concerns and topics have long been central to the field, the two disciplines have exerted little influence on each other until recently. Perhaps this has been because environmental history has concentrated on narratives of massive human intervention in landscapes narratives of the encroachment of the modern world into nature, and African environments have been continually misread as displaying little influence by humans. (2012, 2)

Land is a social construction; it is nature that is under threat. Sometimes, land is a patriotic symbol, and, sometimes, it is a tradable commodity. Even today, it is the nature that takes the blow. It is the highly advanced and civilized people who bomb and violate nature to establish nuclear or military supremacy.

Henceforth, it becomes most essential to understand the impact of postcolonial tendencies on the environment. As a result, there has been already some scholarship in this interdisciplinary area of postcolonial ecologies. Particularly, scholars like Elizabeth DeLoughrey, Robert P. Marzec, and many others have been trying to explain the intersection between postcolonialism and the environment. Robert P. Marzec, in his book, *An Ecological and Postcolonial Study of Literature: Daniel Defoe to Salman Rushdie*, confines his scrutiny to British novels, like, *Robinson Crusoe*, *The Return of the Native*, *Women in Love*, and *A Passage to India*, and Rushdie's works. To be precise, he examines the evolution of colonialism. The book attempts to expose the gradual rise of a discourse of enclosure to reveal the profound impact this discourse has had on humanity and on full-scale colonization.

Marzec discourses about the rise of British Empire and the enclosure movement, which had restructured the entire world. He defines, "An enclosure is the turning of open, communal land into private property" (Marzek 2007, 12). He talks in length about Crusoe syndrome. Robinson Crusoe is unable to set or settle down on the island because he fears it belongs to someone else. However, he slowly adapts, colonizes the island, and begins to rule.

Enclosure act in England has transformed the imperial ideology into material reality. He talks about enclosed land and unclosed land. The unclosed land is associated with colonies like Africa and India. "The relationship between enclosed and unenclosed spaces is more consciously thematized and nuanced in the novels of E.M. Forster" (Marzec 2007, 10). So, colonialism is seen as an act of improving the land by cultivating it and putting it to some use. However, eventually, this use of the land transforms into exploitation as the land is used to benefit the colonizer, instead of helping the colonies.

Deleuze and Guattari propose a term, "stock piling," which Marzec also employs to describe the new economy. He explains how the stockpiling economy of enclosing (the necessity for continual expansion), and the structure of a (non)relation to the other it enforces, is fundamental to the logic of enclosure and crucial to the expansion of British national identity in the colonial world map. He elaborates on the functioning of a profit-driven economy. Marzek cites from both *Kim* and *Heart of Darkness*, bringing examples from colonies across the continents. Marzek also describes and defines the geographical divisions:

> "Territory," for instance, names a juridico-political dispersion of power: a mapped area controlled by a lord, a military commander, an imperial surveyor, a governor, or a nation. "Field" denotes an economico-juridical dispersion: the landlord who encloses a space of land, turning it into his "field" in order to expand his sovereignty and his income (the stock of grain or sheep). "Landscape" indicates a politico-aesthetic dispersion: the comportment of land to an artist's image of organic beauty. (2007, 13)

Graham Huggan and Helen Tiffin begin their book *Postcolonial Ecocriticism: Literature, Animals, and Environment* (2010) with questions like, "Is there any way of reconciling the Northern environmentalisms of the rich (always potentially vainglorious and hypocritical) and the Southern environmentalisms of the poor (often genuinely heroic and authentic)? Is there any way of narrowing the ecological gap between colonizers and colonized, each of them locked into their seemingly incommensurable worlds?" (2010, 6).

Graham Huggan and Helen Tiffin also talk about biocolonialism, biopiracy, environmental racism, zoo criticism, and Val Plumwood's ecofeministic philosophy on nature and imperialism. Graham Huggan and Helen tiffin contemplate the social and cultural hierarchy because the position of marginalized is next to animals. Further, the natives or the tribals are regarded as part of nature and exploited alongside it.

Further, the native culture and nature has also been an exhibit, besides an exploit. The concept of "gaze" also plays a vital role. The colonial gaze and the exotic objects included not just the cultural artifacts but the flora and fauna as well. Now, this is renamed tourism. Graham Huggan and Helen Tiffin quote Wasserman's words, "Indians paraded before royal courts; like turkeys and parrots in cages were the innocent signifiers of an otherness that was . . . exotic, that is, non-systematic, carrying no meaning other than that imposed by the culture to which they were exhibited" (2010, 10).

Graham Huggan and Helen Tiffin remark, "Postcolonial ecocriticism—like several other modes of ecocriticism—performs an advocacy function both in relation to the real world(s) it inhabits and to the imaginary spaces it opens up for contemplation of how the real world might be transformed" (2010, 13). Graham Huggan and Helen Tiffin annotate, "ecology is more than just a network; rather, as the term 'ecology' implies, it is a sphere of co-dependent interaction that connects people to the other ecological beings, both animate and not, that share their phenomenal life-world" (2010, 42).

Elizabeth DeLoughrey and George Handley, in their book *Postcolonial Ecologies: Literatures of the Environment* (2011), draw upon several literary examples to set a comprehensive note. They also use the term "green orientalism" to emphasize the role of empire and imperial politics in ecological transformation over the period of time. They express:

> Certainly postcolonial ecology must engage the complexity of global environmental knowledge's, traditions, and histories in a way that moves far beyond the discourses of modernization theory on the one hand, which relegates the global south to a space of natural poverty, and the discourse of colonial exploitation on the other, which relegates the global south to a place without agency, bereft of complicity or resistance. Our conviction that dominant ecocritical

methodologies that make universalist claims (from the unmarked Anglophone viewpoint of the United States) must address colonial legacies and postcolonial contexts is in con-cert with critics who have cautioned against turning to indigenous and postcolonial ecologies to simply provide moral, spiritual, or financial redemption for the capitalist metropole. (DeLoughrey and Handley 20)

Pro-poor tourism (PPT) is a new and diluted form of colonialism. Anthony Carrigan considers mass tourism as a form of colonialism. "Being able to demonstrate that the currently hegemonic view of a place is historically contingent, political, exploitive, and dependent on its being seen by people as legitimate can be a powerful starting point for a group that lacks economic, political, and institutional power due to years of exploitation" (Carrigan 2011, 84). Erin James cites Rob Nixon's four key chasms between postcolonialism and ecocriticism:

While post-colonialists tend to emphasize hybridity and crossculturalization, eco-critics have favoured discourses of purity, such as narratives of virgin forests or images of untouched wilderness. While post-colonialists often concern themselves with displacement, eco-critics tend to seek out literature of place. While post-colonialists have placed value on the cosmopolitan and the transnational, ecocriticism's origin lies largely in a national, American framework. While post-colonialists have worked to excavate or reimagine the lost marginalized past, eco-critics have leaned towards the pursuit of a timeless, solitary moment of commune with nature. (2012, 63)

Further James adds to this, "To Nixon's four schisms we can add a fifth and sixth: postcolonial scholars tend to be influenced by literary theory, whereas much first-wave ecocriticism was produced in reaction against poststructuralist ideas, and ecocritical pedagogy often takes a place-based approach that relies on visits to the environments students read about—a task difficult to replicate in the teaching of most postcolonial literatures" (2012, 63).

The relevance of postcolonialism and the environment gains even more significance in the second decade of the twenty-first century. The uneven development and post humanity representations across the world call for an academic and literary response.

One may wonder in this context that what is environmental postcolonialism. In short, it is a tool of reading like countless other tools with paradigms peculiar to its own nature. It is an effort to read literature from the standpoint of ex-colonies with their human and nonhuman context. It argues that colonialism was disastrous for both culture and nature. It perpetuates that eco crises of today are the result of Capitalist and Colonial policies. It provides a fertile ground for the wedding between ecocriticism and postcolonialism. It critiques the one-sided profit-driven approach of the colonizers toward their colonies. On universal ground, it scolds human beings for green imperialism across ages and cultures.

The present book is a modest attempt to present environmental postcolonialism in practice. The beauty of the book is that all the chapters are focused on the literature produced in the ex-colonies. Chapter 1 is "Introduction" that provides the conceptual framework for the theory and the book.

Chapter 2 presents a critique of the concept of utopia and dystopia with reference to Amitav Ghosh's *The Hungry Tide*. It argues that in the present scenario, the search for an eco-utopia is inevitable as the world is rendered ugly because of colonial and capitalist policies. Chapter 3 examines the idea of cultural nationalism of Brahminical ideology as imposed upon the sacred groves of Kerala with a view of how one system of thought subordinates in its imperialistic mission the belief system. Chapter 4 exposes how African theater protests against the western tendency to monopolize oil extracted from African reserves and how European powers continued the exploitation of Africa even after the latter gained her independence. Chapter 5 contextualizes the narratives from India's North East with a view to present the natural beauty and diversity of the North East and how human desires to control the surroundings have problematized the cultural and natural landscape. Chapter 6 talks about how Daud Kamal's poetry tries to portray Pakistani landscape in an effort to define the aesthetics of belonging in a postcolonial context. Chapter 7 unfolds the plight of both human and nonhuman in Canada through *Surfacing* by Margaret Atwood with special emphasis on neocolonialism.

Chapter 8 highlights the human tendency to subjugate nature and marginalized communities with reference to *Days and Nights in the Forest* and its cinematic adaptation. Chapter 9 explores the connections between power, dispossession, colonization, and the environment. It looks into the concept of Eco cinema and the problems of anthropocentrism with *Ashani Sangket* as an example.

Chapter 10, via *When I Was Puerto Rican*, discusses the environmental injustices of US occupation/invasion through the lens of settler colonialism theory and demonstrates how settler colonization impacted Puerto Rican livelihood, food, education, and politics, pushing thousands to emigrate in search of better economic opportunities to promote a sense of empathy and understanding. Chapter 11 makes an attempt to analyze a few concerns associated with postcolonial ecology in the writing of the Indian debutant writer and journalist Shubhangi Swarup titled *Latitudes of Longing*. Chapter 12 shows how the Caribbean writers have begun to incorporate technology, such as digital media, not only as a symbol of modernity but, more importantly, as a tool to present the different cultural practices, beliefs, environment, and peoples, creating a collage that allows the readers to experience through different lenses the diverse practices and peoples of the Caribbean Archipelago.

Chapter 13 seeks to explore the idea of nature and its interconnectedness with human life through close readings of J. M. Coetzee's novel, *Life and Times of Michael K* and Chris Abani's novella, *Song for Night*. Both the writers decentralize human beings and situate them as an entity that exists within a network of other entities like animals, plants, nature, rivers, waterbodies, and so forth.

Chapter 14 focuses on *Petals of Blood* by Ngugi Wa Thiong'o and *Roots* by Alex Haley. It illustrates how these novels represent the interconnection between colonialism, capitalism, and industrialization as factors affecting nature and culture in Africa. Chapter 15 examines two of Coetzee's writings—*Disgrace* and *The Lives of Animals*. It argues how Coetzee fluidly moves beyond the culturally encoded boundaries of humans and animals and focuses on the "physiological and bio structural homologies." The chapter tries to situate Coetzee's writings in the larger history of Africa both before and after colonialism. Chapter 16 pleads for a new theory of environmental postcolonialism as the existing theories do not take into account the indigenous perspective. It is an attempt to evolve the aesthetics of environmental postcolonialism from the standpoint of ex-colonies.

BIBLIOGRAPHY

Achebe, Chinua. 2002. *Things Fall Apart*. New York: McGraw-Hill Company.

Carrigan, Anthony. 2011. *Postcolonial Tourism: Literature, Culture, and Environment*. New York: Routledge.

Conrad, Joseph. 2014. *The Heart of Darkness*. New York: Open Book Integrated Media.

DeLoughrey, Elizabeth, and George B. Handley. 2011. *Postcolonial Ecologies: Literatures of the Environment*. Oxford: Oxford University Press.

DeLoughrey, Elizabeth, Jill Didur, and Anthony Carrigan. 2015. *Global Ecologies and Environmental Humanities: Postcolonial Approaches*. New York: Routledge.

Falola, Toyin, and Emily Brownell. 2012. *Landscape, Environment, and Technology in Postcolonial Africa*. New York: Routledge.

Fedar, Helena. 2014. *Ecocriticism and the Idea of Culture: Biology and the Bildungsroman*. Vermont: Ashgate.

Fisher, Michael. 2018. *An Environmental History of India: From Earliest Times to the Twenty-First Century*. Cambridge: Cambridge University Press.

Garrard, Greg. 2004. *Ecocriticism*. 1st Edition. New York: Routledge.

———. 2012. *Teaching Ecocriticism and Green Cultural Studies*. London: Palgrave Macmillan.

Glotfelty, Cheryll, and Harold Fromm, editors. 1996. *The Ecocriticism Reader*. Georgia: University of Georgia Press.
Gokak, Vinayak Krishna. 1970. *The Golden Treasury of Indo-Anglican Poetry*. New Delhi: Sahitya Akademi.
Huggan, Graham, and Helen Tiffin. 2010. *Postcolonial Ecocriticism: Literature, Animals, and Environment*. New York: Routledge.
Iheka, Cajetan. 2018. *Naturalizing Africa: Ecological Violence, Agency, and Postcolonial Resistance in African Literature*. Cambridge: Cambridge University Press.
James, Erin. 2012. "Teaching Postcolonial/Ecocritical Dialogue." In *Teaching Ecocriticism and Green Cultural Studies*, edited by Greg Garrard. London: Palgrave Macmillan.
Malamud, Randy. 2012. *An Introduction to Animals and Visual Culture*. London: Palgrave MacMillan.
Manes, Christopher. 1996. "Nature and Silence." In *The Ecocriticism Reader*, edited by Cheryll Glotfelty and Harold Fromm. Georgia: University of Georgia Press.
Marzec, Robert P. 2007. *An Ecological and Postcolonial Study of Literature: Daniel Defoe to Salman Rushdie*. London: Macmillan.
Monani, Salma, and Joni Adamson. 2017. *Ecocriticism and Indigenous Studies: Conversation from Earth to Cosmos*. New York: Routledge. Paniker, K. Ayyappa. 1991. *Modern Indian Poetry in English*. New Delhi: Sahitya Akademi.
Rodney, Walter. 1981. *How Europe Underdeveloped Africa*. Massachusetts: Howard University Press.
Shukla, Anu, and Rini Dwivedi. 2009. *Ecoaesthetic and Ecocritical Probings*. New Delhi: Sarup Book Publishers Pvt Ltd.
Tharoor, Shashi. 2016. *An Era of Darkness: The British Empire in India*. New Delhi: Aleph Book Company.
Thiong'o Wa, Ngugi. 2003. *Petals of Blood*. London: Penguin.
———. 2012. *Weep Not Child*. London: Penguin.
Walker, Alice. 1984. *Horses Make a Landscape Look More Beautiful*. California: Harcourt Brace.
———. 2010. *The Color Purple*. California: Phoenix.
White, Jr., Lynn. 1996. "The Historical Roots of Our Ecological Crisis." In *The Ecocriticism Reader*, edited by Cheryll Glotfelty and Harold Fromm. Georgia: University of Georgia Press.

Chapter 2

"Transformation Is the Rule of Life"

Environment and the Search for Utopia *in* The Hungry Tide

Suzy Woltmann

> But here, in the tide country, transformation is the rule of life: rivers stray from week to week, and islands are made and unmade in days. In other places forests take centuries, even millennia, to regenerate; but mangroves can recolonize a denuded island in ten to fifteen years. Could it be the very rhythms of the earth were quickened here so that they unfolded at an accelerated pace?
>
> —Amitav Ghosh, *The Hungry Tide* (2004, 198)

Amitav Ghosh's postcolonial novel *The Hungry Tide* (2004) depicts the impact of a devastating tsunami in the Sundarbans Islands. Characters in the novel experience personal, political, and ecological disasters; however, their desire to seek utopia demonstrates a resilience that defies trauma. The search for utopia has generally been implemented in an imperialist way that globalizes, homogenizes, and eradicates difference without allowing consideration for subaltern perceptions of a different utopian ideal. However, in *The Hungry Tide*, the search for utopia subversively allows the subaltern to become engaged in dialogic discourse. In the novel, characters seek utopia and therefore attempt to make home through the lens of aesthetics. Utopia is sought through interaction with the dispossessed (utopia-as-person), attempts to create a utopian society (utopia-as-place), and subaltern death (utopia-as-sacrifice). Utopia-as-person is articulated mostly through Kanai's interactions with the displaced Piya. Attempts at achieving utopia-as-place fall short because of a harsh environmental and political climate, but these projects are not perceived as failures. Rather, the striving for a utopian ideal is worthy of the aesthetic and cultural discourse it creates. Finally, utopian sacrifice and death allow the subaltern to achieve voice. Through these depictions, Ghosh implies that utopia must be the process of living ecologically and being attuned to nature, rather than something one can impose upon a land without regard for its needs and the needs of its inhabitants, both human and nonhuman.

Ghosh's textual corpus deals with issues of postcolonialism, the politics of home, identity, nationality, ecology, environment, and utopianism. *The Hungry Tide* tells the story of the translator Kanai as he meets the cetologist Piya on their way to the Sundarbans. Kanai is traveling to deal with the aftermath of his uncle's long-ago death, while Piya is studying the islands' dolphins. Kanai's uncle, Nirmal, told him stories about Sir Daniel Hamilton, a Scottish colonialist who created a communal settlement on the islands years before. Kanai's old friend Kusum has also died, but her son Fokir works as a fisherman on the island. Consumed by curiosity, Kanai opens a packet that Nirmal left him upon his death, which contains a notebook and letters. One letter asks that Kanai help remember what happened on the island years ago when Nirmal was with Kusum. Piya starts her dolphin studies and meets Fokir, with whom she shares a deep connection despite their lack of shared language. He saves her life when she is about to drown and accepts her offer to take her on his boat to Lusibari, where Kanai is staying with Nilima, his aunt and Nirmal's widow. Kanai begins to read through Nirmal's notebook, which shows how he worked with the community on Morichjhapi and taught children there. Nilima was angry at his involvement because she felt the land should be protected from refugee encampments. However, Nirmal kept providing assistance to the settlers until a final police assault that left him fatally ill. Piya and Kanai go with Fokir to observe the dolphins further, but Nilima begs them not to go due to the extensive environmental dangers. Piya, Kanai, and Fokir witness a violent altercation between villagers and a tiger, after which Fokir explains that the government often values animal lives above human ones. Kanai tries to show Fokir that he is unafraid of the hostile environment but is ultimately frightened enough to leave the trip. When he hears that a cyclone is imminent, he goes back to rescue Piya and Fokir but cannot find them so heads back to Lusibari. He loses Nirmal's notebook in the storm and asks Nilima to allow him to record her version of events along with his memory of those told in the notebook. Piya and Fokir try to find shelter during the storm, but Fokir is hit by a flying branch and dies. Piya later returns to the island to help develop a conservation program in Fokir's honor.

In *The Hungry Tide*, Ghosh provides a fictionalized account of the real Morichjhapi settlement of Bangladesh refugees in the Sundarbans. The Sundarbans are a beautiful but dangerous place, often wreaked by cyclones and monsoons. They are home to Bengal Tigers that kill several hundred people a year. In 1979, the West Bengal government ruled that the Morichjhapi settlement was illegal since it encroached on a reserved tiger habitat. The subsequent evacuation led to the deaths of many refugees, which essentially gave the message that the tiger habitat's sanctity is more important than the subaltern refugee. In his novel, Ghosh examines the tenuous line between real ecological awareness and false environmentalism in the name of economic and political exploitation. He depicts a beautiful topography marred only by insidious, disrupting dangers—human-killing tigers, inter-class tensions, and devastating cyclones. Despite these environmentally devastating terrors, however, the beauty of the area is enhanced by characters' search for utopia. Through the depiction of this search, the novel "seeks a postcolonial ethics and aesthetics that transcends the ideologies of the past, even as it cautiously evaluates the extent to which such a utopian ideal is possible" (Giles 2014, 2). Ghosh seems to argue through his novel that the search for utopia has value in itself; even if utopia itself is impossible to reach, the desire for a better world is not. This desire defies imperialist oppression while producing a rhetoric of the postcolonial sublime, the desire for interconnectivity and understanding between man and nature (1). *The Hungry Tide* encourages a new approach that combines postcolonial theory with anti-imperialist environmental ethics. Since its 2004 publication and Hutch Crossword Book Award for Fiction selection the same year, *The Hungry Tide* has been approached by literary scholars through the lenses of postcolonial, diasporic, Indian English, and island/ecoconsciousness studies. Saswat Das explores Ghosh's metaphorical exploration of home and homelessness that encompass the lived diasporic experience (2006). Pramod Nayar contends that the specter of postcolonial dispossession haunts the novel, resulting in a politically foregrounded uncanny (2010). Nandini Bhattacharya argues for a comparative politics reading of *The Hungry Tide* since its exploration of dualism written in English by an Indian author represents the tradition of "contemporary Indian English writing (embodying its every singularity)" (2013, 59). While Ralph Pordzik (2001), Jessica Namakkal (2012), and others explore the trend of utopian literature as it intersects with postcolonialism to argue for cross-cultural comparisons between these texts, nobody has yet explored the different types of utopian ideals as they relate to the environment in *The Hungry Tide*. Reading ecological value and utopian processes into the text is especially useful today since our planet is more interconnected and cosmopolitan than ever. By returning our attention to a fictionalized version of the problematic relations in the area, both personal and environmental, we can work toward awareness and growth.

The Hungry Tide shows how disturbances in the environment can lead to the search for utopia. As John Su argues, the recent turn toward the aesthetic reverses the anti-colonial contention that aesthetic value is inherently

Western, based on Enlightenment values (2001, 65). However, utopianism in postcolonial literature ensures these values are "reclaimed and redeployed within postcolonial contexts" (66). Utopias and aesthetic function are closely related in Ghosh's works, and he has "consistently portrayed in positive terms his notion of a more egalitarian society" (67). Characters' homes in Ghosh's novels are often ravaged by environmental disaster or political mishaps; however, they (and readers) still experience the aesthetic value of beauty during the search for utopia following these traumas. Utopia allows for the appreciation of aesthetic value while still retaining anti-imperial and postcolonial thinking.

The postcolonial novel looks at identity and interactions in the development of a postcolonial society to formulate a theory of identity politics. Mikhail Bakhtin's theory of aesthetic emphasizing indicates that in novelistic writing, the writer witnesses the interrelationship of the observer and the observed and the moment in which that relationship occurs (1981, 78). The writer steps outside of the relationship to write about it, creating a necessary subjectivity. Further, postcolonial literature is "inherently translated" because it "has always needed to compare and translate among regions, languages, and literatures" (Walkowitz 2015, 169). This inherent translation means that the postcolonial novel provides multiple venues for questions of ecocriticism and environmentalism.

In my analysis of Ghosh's novel, I rely on Bakhtin's notion of transgredience or experiencing life outside of oneself. While anthropocentricism values human supremacy, ecocentrism takes into consideration all the earth's needs. Transgredience encourages authors and critics to view themselves through another's perspective, whether that be human, general nature, or specific beings. It leads to a constant ethical evaluation of one's moral behavior. These other perspectives help one formulate a more moral identity. The author of an ecological book such as *The Hungry Tide* provides a perspective that is both observational and representative of other perspectives. Ecological authors represent being-in-the-world and limitations of human perception quite differently; adopting ecocentrism as an orientation encourages greater eco-comprehension since environmental writing often includes nonhuman others in memory, self-understanding, and perception. Further, the immense diversity of postcolonial literature has a great ability to move audiences because audiences themselves are diverse. Readers can aesthetically empathize through literature by vicariously experiencing other perspectives. Different literary genres have different degrees of distantiation for the reader, but the novel and in particular the postcolonial novel allows for internally persuasive dialogue and therefore altered value systems. In *The Hungry Tide*, these value systems are articulated through the search for utopia-as-person, utopia-as-place, and utopia-as-sacrifice.

UTOPIA-AS-PERSON

One means by which characters in *The Hungry Tide* seek utopia is by locating it in other people. Kanai projects utopia onto Piya, who he sees as a strange amalgamation of man and woman, Indian and foreigner, and outsider and insider. The novel opens with a hunt: Kanai spots Piya "the moment he stepped onto the crowded platform" (Ghosh 2004, 3), and he characterizes her based on her performative identity. Piya's outward appearance is articulated in terms of subversion. She is dressed in androgynous clothing, similar to "those of a teenage boy," has a severely short haircut, stands "like a flyweight boxer" and is free of any culturally dictated feminine adornment (3). Kanai is attracted to Piya's obvious displacement and finds the "neatly composed androgyny of her appearance" to be "almost exotic" (3). Piya thus becomes a source of home for Kanai, as she belongs no-where and to no-place and so becomes the place where anything and all can occur. Kanai finds beauty and home in Piya's diasporic body, the "ideological and symbolic battleground in which foreign-local distinction is played out" (Goh 2012, 342). Yet while Kanai initially views Piya as a liminal space of escape, he eventually falls for her after fearing he will lose her to Fokir.

Piya has a history of not belonging. Her home life was hostile and full of neglect, and Piya was regularly used as an intermediary between her first-generation Indian parents who had never fully assimilated into American society. When she was at school, she was also viewed as an anomaly, as her outward appearance and physical form were different from those of her peers. She must travel constantly for her work, and even her attempts at love have rendered her outcast. Kanai realizes his love for Piya when she asserts that she has decided to "get used to the idea of being on [her] own" (Ghosh 2004, 259). To Kanai, her "true extraordinariness" lies in her displacement (259); he is attracted to this island of a woman who needs no one and nothing and so to him embodies everything.

Instead of responding to Kanai's attempts at seduction, Piya is captivated by the changing behavior of the Orcaella dolphins, theorizing that they are adapting to their ever-mutating environment. She is amazed by the dolphin's cooperation with the local fishermen, with whom the dolphins have a symbiotic working relationship to catch more fish. She asks: "Did there exist anymore remarkable instance of symbiosis between human beings and a population of wild animals?" (168). This belief is further cemented through Fokir's explanation that Kusum had told him that the dolphins are Bon Bibi's messengers. The inclusion of their symbiotic relationship with humans in legend signifies a lasting relationship between dolphin and human—a relationship that seems to contrast with the more antagonistic, and often fatal, relationship between crocodiles, sharks, and tigers with humans in the same environment. Piya's displacement and obsession with something that is not him (the dolphins) make her desirable to Kanai, whose move from superficial utopian attraction moves to a deeper longing for the dispossessed as he interacts with and is changed by the people and environment of the Sundarbans, and in particular by Fokir. Kanai faces his worst fear when he is alone with Fokir: that of a return to his ethnic roots. He learned multiple languages in an attempt to become the Western-watching object instead of the subaltern subject, and the ethnic and racial slurs he yells at Fokir demonstrate instead how very interpellated in the "fraught trajectory of coercive mimeticism" he has become (Chow 2002, 124). Kanai overcomes this by continuing his search for utopia in Piya and transcribing Nirmal's diary in an attempt to open up the subaltern discourse.

UTOPIA-AS-PLACE

While seeking utopia in another person is one means by which individuals respond to environmental chaos, utopian ideology is also represented in the search for a perfect chronotype. In *The Hungry Tide*, a representative of the European West, Sir Daniel Hamilton, attempts to create a utopian settlement of homes in West Bengal that would eliminate racial, ethnic, religious, and class boundaries. This would be a place separate from societal dictates, where preconceptions of status and caste would be replaced by an egalitarian utopia. There would be no "petty little divisions and differences" and "everyone would have to live and work together" (Ghosh 2004, 44). Hamilton buys 10,000 acres of land despite warnings about the Sundarbans being home to a terrifying climate, dangerous animals, and political strife. There is "no prettiness" in the treacherous location, and yet it is known as "the beautiful forest" (8). These paradoxical descriptions show how a hazardous environment can be perceived in vastly different ways depending on one's ecological awareness. Hamilton hopes to "build a new society, a new kind of country . . . run by cooperatives" that could be "a model for all of India" where people could live together without hegemonically imposed social caste distinctions (45).

Hamilton's dream of a utopian home is untranslatable in the harsh Sundarbans environment; however, and after his death everything reverts once again to imposed regimes, marked social difference, poverty, and subalternity. These political and biological realities dictate the means by which characters in *The Hungry Tide* interact and are involved with one another. Fokir protects Piya from animal attack and is ultimately killed in a giant storm; Kusum's father is killed by a tiger, altering the course of her existence; Kanai's experience with the land changes his very being and allows him to fall in love; and Nirmal becomes involved in refugee politics, which are then altered by the dire landscape. Indeed, the "landscape of the Sundarbans becomes a problematic war-field between forces of nature and that of the human world" (Rath 2010, 18–19). Based on their attempts to understand their environment, some characters are less at war than others, and yet ultimately, they all must answer to nature. Hamilton's conception of utopia is essentially lost in translation, as his imperialist discourse and Western implementations surrender to the biological dangers and political strife of a treacherous environment.

Like Hamilton, Nirmal also dreams of a utopian homeland, but as a man "possessed more by words than by politics" he is unable to reconstruct his words and desires as something tangible (Ghosh 2004, 282). Much like Kanai's feelings for Piya are tied to how she behaves with the environment and Fokir, Nirmal's interest in Morichjhapiis provoked through his extramarital desire for Kusum. When Nirmal initially hears of the "tens of thousands of settlers" traveling to the island belonging to the Forest Department, he responds with apathy, as it is "no business" of his (132). His discussion with Kusum alters this attitude. In transcribing the story of how she came to Morichjhapi, he fancies her the representative of these dispossessed people and himself their historian. Nirmal emplots the tale of Morichjhapiand its refugees as that of a doomed romance and insinuates himself inside its history, though in actuality, he has very little to do with the politics and cultural happenings of the land. By using Kusum as a prism through which to view Morichjhapi, he places her inside a utopian ideal, which may never be achieved but through death. The Morichjhapi refugees themselves exist in a diasporic state of no-place or non-place, though their living places are oftentimes not ideal. The idea of home for these refugees is instead something purely in the emotive terrain; it becomes something they strive for but may never achieve. Non-places are represented through refugees and their "memories, recollections, blurring of lines between narrative time and real historical time, and the idea of 'places' as a desire and a process for the gratification of that desire, are the aspects of these 'non-places' " (Rath 2010, 30). The utopian ideal is omnipresent in the mind of the migrant refugee and exists as a point temporally distant from their spatial reality, ensuring it is a time-space construct that may never be attained. The place they reside may also be classified as a non-place, for it is a place of struggle, possible displacement, and apprehension of government and mobility.

Nirmal becomes obsessed with the idea of Morichjhapi, which he believes can be a communist, environmentalist refuge. He views the island as an "egalitarian world disentangled from capitalist exploitation" (Su 2001, 39). In this intentional community of no-place, he finds "an astonishing spectacle—as though an entire civilization had sprouted suddenly in the mud" (Ghosh 2004, 159). Nirmal romanticizes this notion of creating utopia out of nothing and believes this land would be a crowning achievement of diasporic humanity; a land whose significance would "extend far beyond the island itself" as "a safe haven, a place of true freedom for the country's most oppressed" (159). However, Nirmal's hope proves to be dangerously idealistic. He and Nilima are shocked to discover that the environmental hazards of the island do not allow their dreams to take hold. The soil is salty and will not bear crops; storms cause periodic flooding; and crocodiles, sharks, snakes, and tigers kill settlers daily (79). Nirmal and Nilima "had not expected a utopia, but neither had they expected such destitution. Faced with this situation, they saw what it really meant to ask a question such as 'What is to be done?' " (79). This gap between discourse and material realities lead to a dangerous conclusion wherein the most hopeful, poetic ideas could not provide food and goods, and establishing a representative discourse could not alter the economic establishment of an oppressed minority. Nirmal discovers that discourse does not exist in a separate realm but instead is engaged in constant conversation with material existence.

Because of his insistence on romanticizing what is in actuality an extremely hazardous environment, Nirmal's diary depictions are especially problematic. The diary is written at a place spatially and temporally distant from the present, and he writes it as an outsider. Nirmal is an intellectual, a theorizer, and a poet, who attempts to transcribe the political happenings of a group of refugees of which he is not a part. The novel connects the present of Piya, Kanai, and Fakir's experiences with the past of Nirmal's diary. In his diary, Nirmal relies on Euro-colonial influence. He quotes Rilke's poetry and idealizes Sir Daniel Hamilton's utopian project while willfully ignoring the environmental dangers faced by the settlement. This implies that colonial influence is inherently tied to ecological awareness. Indeed, "in *The Hungry Tide*, nature is interwoven with history, and the rhythm of life here indicates a perennial struggle between ebbs and flows, subjects and the environment, self and other, humans and animals, water and land, life and death, history and nature, fresh water and salt water. Everything is open, always in the process of becoming" (Chen-Hsing Tsai 2017, 152). Nirmal wants a perfect place and in his inscription of the settlement neglects to consider that nature is always in the process of becoming.

Nirmal is dead at the time the diary is read, and so his words gain profound importance of time and space separation and voice through death. According to Bakhtin, the authoritative discourse has power because it seems so far removed from its audience; it emerges from no-place, no-time, and omnipotence. Alternatively, internally persuasive discourse is "half ours and half someone else's" (Bakhtin 1981, 582). Because Nirmal's diary was written so long ago and since he is now passed, his words have authoritative power. His incorporation of the Bon Bibi legend also grants the diary power through mysticism and mythology. Bon Bibi, the protector of the forest, is supposed to also protect humans—if they act in the environment's best interests. She is "the forest's protectress" (Ghosh 2004, 304), who "rules over the jungle ... the tigers, crocodiles and other animals do her bidding" (96). By including her legend in his diary, Nirmal implies that his diary should have the same authority and environmental significance as Bon Bibi. This implication works: when Kanai first reads the diary, he interprets it as historical fact, forgetting that history, too, is plotted. Nirmal's impulse to romanticize his experiences into a story neglects the harsh reality of the Morichjhapi environment and so denies the complexities of what actually happened at Sir Daniel Hamilton's commune.

Unlike her husband, Nilima understands the limits of utopian idealism. Rather than yearning for revolution or searching fruitlessly for a home that may never be translatable except through death like her husband Nirmal, Nilima shows that adapting outside technology and seeking external knowledge may help people to respond to environmental hazards. In particular, these avenues may help mediate the subaltern experience. Globalization generally conveys Western, hegemonic ideologies that may translate as encoded imperialism. However, these ideologies can be appropriated by those impacted in order to benefit the community. Nilima founds and runs the Badabon trust, which utilizes equipment and technology from both local and foreign sources in order to provide aid and services to locals. Nilima's persistence in running the trust, a nonprofit organization "widely cited as a model for NGOs working in rural India," initially confuses Kanai, who cannot imagine giving up a life of "creature comforts" to dedicate one's life to the "dire poverty of the tide country" (Ghosh 2004, 29). However, while he cannot envision a similar path for himself, he respects her dedication to such a seemingly impossible location and community. Similar to Nilima, Fokir's wife Moyna seeks out education and trains to be a nurse in order to provide assistance to subalterns. However, her understanding of Western ideology causes her to look down on her husband and people like him, who try to remain unaffected by globalization efforts. Both Nilima and Moyna deny utopian idealism but do integrate globalized cultural values, ideology, and technology in order to better the place and time they exist in and still retain hope for a brighter, if not perfect, future.

UTOPIA-AS-SACRIFICE

Alternatively, Fokir tries to resist modernization and globalization and represents the impossibility of subaltern voice except through the utopian state of silence, sacrifice, and death. Fokir's physical appearance is described in terms of alteration and a potent relationship to his homeland. He has "the grizzled look of an experienced hand," his clothes are sparse and have a utilitarian purpose, and his body shows his time spent "slowly yielding his flesh to the wind and the sun" (Ghosh 2004, 36). When Piya first sees Fokir from afar, she thinks he is an old man, which demonstrates his connection to an ancient earth and way of life; however, upon closer view, she is startled by the aesthetic value of his youth and attractiveness. Fokir is "not wasted but very lean" with limbs "almost fleshless in their muscularity" (40), and his stance invokes both destitution but also defiance (41). This biological description makes him seem animalistic, which is befitting since he is the closest to the environment of any of the characters; he lives "in an organic fusion with the ecosystem of Sunderbans" (Rath 2010, 24). This organic fusion is initially proven when Fokir is the one person who can take Piya to the dolphins, and again when he ultimately dies in an epic battle between man and nature. Fokir is also an aesthetic figure because he believes "people must safeguard spheres of existence outside of politics" (Su 2001, 70). Whatever other people do, Fokir "does just the opposite" (Ghosh 2004, 129); he has no interest in the world outside of his personal relationship with the land and water, which is "enough" for him (263). As an aesthetic figure, Fokir demonstrates a natural beauty untouched by modernity or capitalist appropriation, and thus "provides the basis for an allegory of the limits of bourgeois society's capacity to extinguish alternatives to itself" (Su 2001, 70). He represents a utopian ideal because he shows that there remains part of existence untouched by the imposing of pseudo-universals.

Fokir's existence as separate from homogenic structure and discourse necessitates his death. He becomes Piya's organic connection to the Sundarbans after he sacrifices his life for her; it is through him that she establishes an ecological existence in the harsh terrain. When they can tell a dangerous storm is coming, Fokir does everything he can to save Piya. In trying to secure the boat, his face and chest become "crosshatched" with cuts (Ghosh 2004, 323). He and Piya share a magical encounter during the eye of the storm, when they witness a tiger swimming past them to the island's thickets. Despite her protests, Fokir shields Piya from the storm with his own body, and this moment is nearly transcendent. Their bodies "were so close, so finely merged that she could feel the impact of everything hitting him, she could sense the blows raining down on his back. She could feel the bones of his cheeks as if they had been superimposed on her own; it was as if the storm had given them what life could not; it had fused them together and made them one" (333). This signifies that Fokir and Piya can only truly come together when they also come together with nature—but that this meeting necessitates sacrifice: at some point during this melding, Fokir dies.

Since he resists interpellation, Fokir is " 'an ideal figure,' a utopian concept designating the limits of hegemonic thought" (Li 2009, 1). The ideal figure is unsustainable, however, and so utopia is discovered instead through sacrifice. Through death, the subaltern achieves voice as a demonstration of alterity. It is only after Fokir's death that the other characters are brought to action. His death "provides the occasion for a meditation on subalternity as a critical alternative to dominant regimes of power" (1), which grants the ability to view the world in another, utopian way. Fokir must die in order to "serve as an irreducible idea" (1). Since any true utopia may only be achieved in another realm, in another place, death provides the means to access this alternative existence.

As Edward Said contends, the non-West has been constructed in the Western imagination through a combination of ideological apparatuses that presents the Orient as Other and inferior (1979). These apparatuses create a system that silences non-Western voices. Subaltern voice must be translated through Western academic discourse in order to be heard, but it is paradoxically this very translation that creates a misrepresentation of the subaltern, as "language necessarily involves an act of violence whereby the object of representation is defined exclusively in terms of the subject's categories of understanding" (Su 2001, 39). Thus, subaltern voice often disappears in the process of translation to Western discourse (Spivak 1999, 190). There may be no accurate representation of the subaltern, as it becomes lost in translation. However, through silence and through death, the subaltern achieves a voice that needs no translation. The subaltern may provide perspective through dialogic discussion and heteroglossic voice as represented through literature and conceptions of home. The impossibility of subaltern voice "results in the paradoxical condition in which home and death are linked, in which the subaltern's death or disappearance enables the subaltern to fulfil the ideal role of the resistant and in appropriable other" (Li 2009, 2). The subaltern voice is loudest in its silence when it is "the unemphatic agent of withholding in the text" (Spivak 1999, 190). It is this power in withholding that creates the subaltern home, a no-place that is untouched by colonial discourse and power regimes. It is only through Fokir's death that "makes possible subaltern inaccessibility, unfigurability, and singularity" while also creating a "utopian alternative to the postcolonial present" (Li 2009, 3). Through death, Fokir defies the attempts of others to translate and modernize the simplistic fisherman. It is defiance through death, agency through silence, and subversion of imposed ideology through sacrifice. Kusum's death foreshadows that of her son, and his death is a return to his mother and her silent subalternity. Both subalterns die so that their death and silence may create a world with more potentiality for utopia.

CONCLUSION

The Hungry Tide explores, dramatizes, and problematizes the relationship between human and nonhuman claims on environment, especially in a postcolonial context. The novel takes an ecocritical stance and complicates the issue of looking at man and nature as separate spheres with one having nonnegotiable dominion over the other. This separation encourages the search for a romanticized utopian ideal in nature, which often leads to skewed and dangerous environmental policies. When nature is romanticized and viewed as an untouchable object, it becomes just as Othered as when humans attempt to establish dominion over nature without regard for symbiotic systems. The novel shows how the separation of culture from nature leads to ecocritical issues and that reaching for a false utopia by Othering—either by making other people Others or by making the environment an Other—leads to something that may only be concluded through destruction of environment and self. However, when utopia is seen as a process of ecological soundness and symbiotic relationships, ecocritical awareness and true utopia may be achieved.

Characters in the novel are constantly seeking a utopian ideal as a means of escape from the present or in hope of a better future. These ideals exist largely within the "terrains of the mind" and may be translated through material existence (Rath 2010, 14), found in representations of others, or exist as idealistic endeavors toward a better world. The interrelatedness of temporal and spatial realities deconstructs and subvert the material and cultural binary, and through aesthetic portrayals of man, nature, and death, utopia is sought for and sometimes achieved (15). In *The Hungry Tide*, utopia is sought in outside perception of the dispossessed, in the various ways individuals have attempted to construct it as a tangible place, and through subaltern sacrifice. These utopias are translated through hope, idealism, and the search for improvement. Applying these vectors to other forms of communication encourages subaltern voice and an ecoconscious ideal. While *The Hungry Tide* was published fifteen years ago, its exploration of postcolonial ethics and environmentalism is even more significant today. Recent ethnographic studies show that the Sundarbans are suffering from climate change, which would be worsened by a proposed coal-burning Rampal Power Station. Advancements in technology have increased the disparity between subaltern citizens and the local elite. It is vital that we turn the lens to focus on postcolonial literature and its aesthetic representations in order to strive for a more egalitarian approach to these crises and to allow ecological symbioses. Ghosh calls for a world in which this discourse is translated into material existence, and there is a more socially conscionable way of life in which cultural heterogeneity is uncompromised. In this way of life, people would be open to expanding their horizons, reexamining prejudices, and recognizing their own place in the world and in relation to postcolonialism, globalization, and the subaltern.

BIBLIOGRAPHY

Ambethkar, Raja, and Jaya Raj. 2012. "Restoration of Human Spirit in *The Hungry Tide* of Amitav Ghosh." *The Criterion* 3, no. 3: 1–10.
Bakhtin, Mikhail Mikhalovich. 1981. *The Dialogic Imagination*. Austin: University of Texas Press.
Bhabha, Homi K. 1994. *The Location of Culture*. London: Routledge.
Bhattacharya, Nandini. 2013. "Revisiting Amitav Ghosh's The Hungry Tide: The Islam/English Dynamic." In *Writing India Anew: Indian English Fiction 2000–2010*, edited by Krishna Sen and Rituparna Roy, 59–74. Amsterdam: Amsterdam University Press.
Chen-Hsing Tsai, Robin. 2017. "Animality, Biopolitics, and Umwelt in Amitav Ghosh's the Hungry Tide." In *Animalities*, edited by Michael Lundblad, 148–167. Edinburgh: Edinburgh University Press.
Chow, Rey. 2002. *The Protestant Ethnic and the Spirit of Capitalism*. New York: Columbia University Press.
Das, Saswat S. 2006. "Home and Homelessness in 'The Hungry Tide': A Discourse Unmade." *Indian Literature* 50, no. 5: 179–185.
Ghosh, Amitav. 2004. *The Hungry Tide*. London: HarperCollins.
Giles, Jana Maria. 2014. "Can the Sublime Be Postcolonial? Aesthetics, Politics, and Environment in Amitav Ghosh's The Hungry Tide." *The Cambridge Journal of Postcolonial Literary Inquiry* 1, no. 2: 1–20.
Goh, Robbie B. 2012. "The Overseas Indian and the Political Economy of the Body in Aravind Adiga's *The White Tiger* and Amitav Ghosh's *The Hungry Tide*." *The Journal of Commonwealth Literature* 47, no. 3: 341–356.
Li, Victor. 2009. "Necroidealism, or the Subaltern's Sacrificial Death." *Interventions* 11, no. 3: 275–292.
Namakkal, Jessica. 2012. "European Dreams, Tamil Land: Auroville and the Paradox of a Postcolonial Utopia." *Journal for the Study of Radicalism* 6, no. 1: 59–88.
Nayar, Pramod K. 2010. "The Postcolonial Uncanny: The Politics of Dispossession in Amitav Ghosh's *The Hungry Tide*." *College Literature* 37, no. 4: 88–119.
Pordzik, Ralph. 2001. *The Quest for Postcolonial Utopia: A Comparative Introduction to the Utopian Novel in New English Literatures*. New York: Lang.
Rath, Arnapurna, and Milind Malshe. 2010. "Chronotypes of 'Places' and 'Non-places': Ecopoetics of Amitav Ghosh's *The Hungry Tide*." *Asiatic* 4, no. 2: 14–33.
Said, Edward. 1979. *Orientalism*. New York: Vintage.
Spivak, Gayatri Chakravorty. 1999. *A Critique of Postcolonial Reason: Toward a History of the Vanishing Present*. Cambridge: Harvard University Press.
Su, John. 2001. "Amitav Ghosh and the Aesthetic Turn in Postcolonial Studies." *Journal of Modern Literature* 34, no. 3: 65–86.
Walkowitz, Rebecca. 2015. *Born Translated*. New York: Columbia University Press.

Chapter 3

Cultural Nationalism and Sacred Groves of Kerala

Anupama Nayar

This chapter will analyze the position of the sacred groves or *kaavus* of Kerala in postcolonial India, located ironically in the interstices of neocolonial ideologies of neoliberal economics, religious secularism, political and militant cultural nationalism, changing Malayalee ecological consciousness, and much needed natural resources management.

> Tension between continuity and rupture, mimesis and repudiation, lived hybridity and the quest for authenticity, is a central condition of post colonialism. It has become the subject of much scholarly inquiry into colonialism and its consequences in India and other countries in Asia and Africa. Disciplinary divides have separated these inquiries into questions concerning institutions (including state formation) and identities (including subject formation). A fruitful synthesis emerges when both sets of questions jointly animate inquiry, and often it is in the realm of environmental history that a new materialism has interwoven with the study of ideas and representation in the examination of postcolonial predicaments. (Sivaramakrishnan 2003, 3)

Says K. Sivaramakrishnan in his article, "Nationalisms and the Writing of Environmental Histories." The varied discourses, converging at some point and sometimes diverging to the point of negation and self-contradiction would require operational definitions of some concepts to clearly delineate the subtleties and nuanced understandings presented in the arguments. The concept of the sacred grove or the *kaavu* will be understood as in the words of M. Gadgil and V. D. Vartak, "Sacred Groves are tracts of the most valuable of legacies from the primitive practices of nature conservation." They are "tracts of sacred forests which have been completely or nearly completely immune from human interference on grounds of religious beliefs." "A Sacred Grove is a patch of vegetation ranging in extent from a few trees to forty hectares or more, which is left undisturbed because of its association with some deity. In its original form, this protection forbade any interference with the biota of the grove whatever, and not even leaf litter was removed from it, nor was grazing or hunting permitted within the grove. Even when the protection has become less stringent, any removal of live wood continues to be taboo. The groves therefore represent a sample of the vegetation in its climax state."

Cultural nationalism is foremost to be understood as a political imposition and its objective is to erase cultural heterogeneity of the Indian subcontinent and the plurality of identities, the nation state claims. "The citizens of a multi-national country have to often distinguish between their pan national identity and their individualistic identities. In such cases, Edward Shills observes that the people hold on to both civil-political and primordial ties at the same time. A disturbance in this framework of dual ties leads to a socio-political and cultural conflict with a feeling of suppression amongst its citizens with the singular national identity being questioned and critiqued" (Athreya 2016, 3). Colonialism played a big role in shaping the history of most pluralist nations that are located in the African, Asian, and the Latin American continents. Through the policy of divide and rule, the European colonialists created a sense of hostility and confusion between two or more dominant religious communities existing in the colony. In India, this is evidenced in the fact that the Hindus and Muslims became increasingly polarized during colonization despite coexisting harmoniously for centuries before colonization. This psychological drifting led to a constant feeling of neglect within both communities. At a macro level, pan India nationalism was seen as being anti-colonial with dual objectives of getting rid of oppressive colonizers and establishing a sovereign republic. However, at the provincial level, nationalism was a form of cultural consciousness that aimed to protect different cultural communities in their homeland. Nationalism in India acquired a particular sociopolitical connotation at both the micro and the macro level. Unlike Europe, India with many distinct nationalities did not proceed to create independent republics instead, the states preferred to retain their cultural identity within a larger sovereign and political nation state framework. It is therefore apt to summarize that cultural nationalism in India operates mostly within the cultural framework of national identity and political framework of autonomy posited in the federal states. Globalization impacted classical nationalistic sentiments for it has affected people and nation states, not only economically and politically, but also socially and culturally. A grave outcome of globalization is that, it has created an unequal world in terms of resources making migration necessary. Migration became important as a means for seeking better economic opportunities in a more developed economy than one's home nation. The sudden increase in the rate of migration post globalization has subsequently led to a rise in xenophobic sentiments, gradually pulling societies into an age of anxiety. A natural offshoot of this anxiety is increasing fear and suspicion among the people of different cultures and ethnicities because of lack of knowledge of that culture or the inability to understand them. Nationalist thinkers therefore fear the loss of their culture because of the influence of foreign cultures. In contemporary times, multicultural societies foster nationalism differently. In most democratic societies, minorities are protected by the law, and the state takes them into consideration while formulating federal policies. "In India, globalization has played an important role in the emergence of Hindi Cultural Nationalism propagated by right wing nationalists. According to Appadurai, Hindu Nationalism can be seen as a middle class, high caste project of cultural homogenization. The aim is to create a unified and homogenized Hindu political entity. Hindu nationalists maintain that the word of the majority community should prevail over the others in a modern democratic state" (Athreya 2016, 4). Hindu majoritarianism in India rests on two basic assumptions: one that the nation states can be built successfully only if there is a shared cultural identity. The other that Hinduism is not just a religion but also a way of life. Thus, by secularizing Hinduism, cultural nationalists claim that this particular brand of Hinduism is the shared identity of the entire population, no matter which religion they belonged to. Not surprisingly, this rise of a radical form of Hindu nationalism directly coincided with India's assimilation into global systems of production and consumption. Economic liberalization in India led to an alarming situation of extreme wealth and poverty. This undoubtedly had an important role in creating communal tensions as it reinforced a seemingly felt religious divide. Cultural nationalism in pluralist societies such as India stem from economic disparity in the population. Cultural or ethno-linguistics identity in postcolonial India is seen mainly as a political tool for better economic and political representations. The state of Kerala, often referred to as "God's Own Country" worldwide is a unique geographical entity. Even before the days of European colonization, the Indian subcontinent was a fabled land arduously documented by travelers from Phoenicia, China, Arabia, South East Asia and Rome. All through the centuries of foreign trade and conquests, the one region in the *Indies* that had allured them was the land of the black gold Kerala, with its diverse forms of cultural and natural wealth casting a hypnotic spell on the foreign explorers.

> Kerala, located at the southern tip of the Indian subcontinent, is noted for its eminent contribution to the country's intellectual and cultural landscape. In the poetic language of Mahakavi Vallathol Narayana Menon, Mother Kerala "sleeps with her head on the lap of the Sahyadri clad in green" and "her feet pillowed on the crystal ocean sand, Kumari at one end and the Lord of Gokarna on the other." This palm-fringed paradise is 'a mystery inside a riddle inside

an enigma' to many. It is a land where nature still holds her own diverse cultural influx and assimilation which have been successful in forging a common ethos and finds a common identity and cascades delicately down the hills to the golden coasts and is lined by lush green coconut groves. The sea which gave birth to Kerala also helped in formulating her history. Peaceful interaction with far-flung lands through trade had built up a tradition of more than two millennia before the incursions from modern Europe symbolized by the landing of Vasco Da Gama at Calicut in 1498, changed the temper of the contact, ". . . loaded it with conflicts and inflicted a turbulent phase of history which ended only with independence." (Kumar Sudarsana 2018, 65)

THE SACRED GROVES

The concept of the sacred groves is not particular to India. From time immemorial all over the world, sacred groves have existed as patches of densely wooded areas, containing unique flora and fauna with perennial water sources in the vicinity. Many sacred groves have been preserved as sustainable resources, ensuring the conservation of valuable gene pool and as a first major effort to recognize and conserve biodiversity. The global environmental crisis penetrated into every aspect of human kind's existence: physical, social, economic, political, cultural, psychological, and spiritual. Consequently, people who attempted to deal with this crisis used multiple epistemic approaches, including the natural and the social sciences, ethics, and religion. Historically, conservation had its roots in pre-agricultural societies. It was often in the name of the divine and the fear of the sacred, that natural areas such as forests, trees, animals, water bodies, and grasslands were preserved by ancient societies. The sacred groves of Europe are part of history and legends and are conserved as relics. However, "they are a living tradition especially in India" (Hughes and Chandran 1998).

> In ancient (pre-Christian) Mediterranean region patches of forests were preserved in the name of gods. Viewing such a stately patch of woods, the first century BC Roman poet Ovid said, 'Here stands a silent grove black with the shade of oaks; at the sight of it anyone could say, "There is a god in here!" '
> The sacred groves of the Mediterranean as well as of India had distinct borders. Tree felling, collection of biomasses, removal of earth, hunting, fishing, farming, grazing of domestic animals, and use for residences or other buildings was forbidden. Specific rules varied from grove to grove, but in general the biodiversity was protected. Exceptions were allowed in times of need. (Hughes and Chandran 1997, 413–427)

The *kaavus* of Kerala—Vestiges of a Buddhist Tradition Kerala—has been an integral part of the Indian nation state; its history is part of the general history of India and its unique culture has enriched the composite culture that India has always boasted of. "The geographical location of Kerala as a narrow strip of land hemmed between the Western Ghats on the one side and the Arabian Sea on the other has had considerable impact on the course of its history" (Menon 2008, 10). The historian A Sreedhara Menon also notes in his book, *A Survey of Kerala History* (2008), that the state enjoyed "insularity" and gave "it immunity from the convulsions which shook Northern India" referring to the foreign invasions that took place from across the Northern borders. The advantage of this geographical insularity was that it took longer time for Brahmanism, Buddhism, and Jainism to enter Kerala. Therefore, Kerala was able to evolve its own way of life and social institutions, unhampered by external interferences and influences. This led to the growth of some peculiar traditions and social institutions like the *Marumakkathayam* or the matrilineal inheritance system and polyandry unique to Kerala. Kerala also developed its own distinct style of architecture and arts. Kerala also came to be known for its rich and abundant verdure.

> Kerala is known for its profuse and green vegetation. It has also been known as the land of magic. The richness in green and magic is one of the peculiarities of Kerala landscape, which is covered with networks of kaavus.... A kaavu is a grove consisting of several sacred and fearful trees, demigoddesses and demigods, ancestors, artefacts used in rituals, ritual actions, narratives of the superhuman power (shakti) and fearful fault (doosham) of the kaavu and place. Size and age of kaavus vary. There are big, small, old and new kaavus. Some are rich in biodiversity, while some kaavus have only a few trees. Divinities in kaavus also vary but there usually are ancestors, ghosts, demigoddesses, demigods, and the divine snake Naga in them. Malayalees say that there are life-force (shakti) and fault (doosham) in kaavus; liminal places connecting the social and the superhuman, enabling the transformation of life into place and death into life. (Uchiyamada 2001, 107–141)

However, historians believe that the concept of the *kaavu* as a sacred grove would have originated with the advent of Buddhism to Kerala during the Sangam Age, though the state enjoyed luxuriant vegetation and the earliest Dravidian societies revered select green patches in the forest. The sacred groves in Kerala are locally known as *Ayyappan kaavu* or *Sasthankaavu*, *Bhagavathikaavu* or *Amman kaavu*, *Vanadevatha* and *Cheema*, or *Cheerumba* and *Sarpa Kaavu* depending upon the ownership and the deities to whom these groves are dedicated.

In a blog entry by Ajay Shekar on the "Panayannar Kavu: The Sacred Grove by the Pampa near Parumala and Niranam," he writes about the still noticeable Buddhist artifacts, influences in rituals, and place names of various kaavus in Kerala. He records, "Now it (Panayannar Kaavu) is a Kali temple but it is clear from the name and the surviving diverse vegetation that it was an ancient Sangha Arama or Buddhist sacred grove by the Pampa before the early middle ages. The very word Kaavu is from Kanyakavu or Kanyastree the Buddhist nun; as it was the nuns and monks who nurtured the medicinal natural grove around their shrines. The Kaavu culture in Kerala is a reminiscence of the conservationist culture that originated with the Asokan missionaries in BC third century" (https://ajaysekher.net/tag/buddhism-in-kerala/).

M. D. Subash Chandran in his well-argued and expository article titled "From the Shadows of Legitimacy Problems and Prospects of Folk Healing in India" (2016) asserts in the context of the origin of Ayurveda in Kerala:

> South India came under Buddhist influence from the time of Ashoka (3rd century BCE) with the visit of Buddhist monks (arhats). Buddha Dharma was propagated through monasteries and learning centers and medical services were rendered from the monasteries by knowledgeable monks. Buddhism found wide acceptability among the masses as it opposed the oppressive caste system of Hindus, as is evident from the Sangam Tamil works of early centuries CE. Its decline from 7th century started due to various reasons which weakened the Buddhist Sanghas; the major reason was the revival of Brahminism. However, despite its weakened state Buddhism lingered in the south until the 14th century. As far as Kerala is concerned, many subscribes to the view that several temples bearing the name Sattan-kavu and Aiyappan-kovil etc. which exist to this day, were former Buddhist shrines. Sattan (colloquial of Sastha) was a name for Buddha and kavu refers to a garden or a monastery. Hence Sattan-kavu refers to a monastery of Buddha. It is well known today that the kavus of Kerala are worship places. Kavus are exclusively sacred groves, or sacred grove with small shrines or temples, or temples which have lost their groves, as many are today (Chandran 2016, 6).

Thus, it can be safely concluded that what is in contemporary times seen as a major high caste Hindu cultural artifact in Kerala had its origin in early folk and medieval Buddhist traditions.

THE PRECOLONIAL *KAAVUS* OF KERALA

In precolonial Kerala, just as much in other parts of what was known as India, "extensive tracts of land were controlled by local communities and used in a sustainable fashion. The thousands of endogamous caste groups of the Indian society had very diversified patterns of resource use and were at the same time linked in a web of reciprocity. Such organization favoured sustainable use of forest resources. The communities enforced strict protection of some forest patches, which apart from sacred groves included also village forests with regulated harvests. The regulations included restriction of seasons for harvests such as leaves for manure, collection of non-timber products as well as family-wise limit for harvests, firewood for instance" (Chandran and Ramachandra 2006, 6). These are later understandings of the ancient societies and their polity as discussed by anthropologists, sociologists, and historians in documented history. Since India privileged oral to written transmission of knowledge, ancient worldviews were encompassed in legends and lore. Hence, this chapter will take the stance that precolonial understanding of the *kaavus* of Kerala is found in the legends and lore of Kerala compiled in the eight-volume *Aithihyamala* (1909) written by Kottararathil Sankunni. *Aithihyamala* originally written in Malayalam in the twentieth century are narratives compiled from texts, which were in circulation for over a long indeterminable period of time in the oral tradition in the Malayalam-speaking region now known as Kerala. The myths and legends in *Aithihyamala* recount the *utpattikatha* or origin myths, *mahatmya* or a story about its greatness, and *sthalapurana*, which depict traditions that have been nurtured, preserved, and circulated of particular *kaavus*, which in Malayalam means not only the sacred grove but also a temple and its locale. Though written during the colonial times, the first few volumes of *Aithihyamala* "texts maintained a position of their own as a narrative genre and preserved a regional orientation by their refusal to passively meld into dominant Hindu narrative and ritual traditions" (Binny 2016, 4). This chapter will not look at *Aithihyamala* as a source of history or as historical evidence but as narratives in informing and complementing historical documentation and providing counter and alternative possibilities of emplotment. A sense of the past is not always internalized through formal history, but by means of mythology, legends, popular beliefs, customs, and rituals. Thus, the legends and myths can be specific kind of reading along with history to understand the indeterminacy of a particular period of time, actual or imagined, as lived experience and fantasy. Mythology here is viewed as a primeval attempt to capture memory, history, and life-worlds of a people at a particular timeline in history.

Some of the stories in *Aithihyamala* revolve around the characters of the benign *Devi* or the goddess and the demonic *Yakshi* or the shape shifting female spirit guarding their kaavus and punishing the trespassers. These two manifestations of the feminine, the anthropomorphized goddess and the *Yakshi* reflect the social as well as the ritual status of women in those times and societies. Temples and *kaavus* of the goddess and the *Yakshi* covertly played a major role in the economy of the region. The diabolic *Yakshi* in *Aithihyamala* symbolized unbridled female sexual desire and is quite different from the figure of the tree nymph called *Yakshi* in temple iconography and certain Sanskrit texts. The *Yakshi* is portrayed as a beautiful woman who allures unsuspecting men to her dwelling, the dark palm tree (*karimpana*, a tree usually found in sacred groves and is not felled), which would be transformed to a palace by her sorcery for the bewitched victim. She fulfills the man's carnal desire, but takes his life in return. She devours the man's flesh and blood after the amorous union, littering the ground below the palm tree with his nails and hair. This, in turn, sprouts into the blood-colored Ixora flower bush commonly found in the Kerala landscape. The narrator of *Aithihyamala* suggests that the *Yakshi*'s seductive and fatal hold on the victim can be released by the powers of the goddess. In one such story, the *Yakshi*'s attempt to kill a Brahmin is thwarted by the *devi* whose powers save him. The *devi* and *Yakshi* are both beautiful and powerful. While the former is predominantly good and chaste, the latter is predominantly vengeful and murderous. The moral anxieties of the cultures wherein the *devi* and *Yakshi* stories were prevalent usually assigned a superior status to the goddess. The *Yakshi* is either annihilated or disciplined to fit into conventionally normative modes of womanliness. A repeating motif in the *Yakshi* stories is the Brahmin man's escape from the clutches of the *Yakshi* with the help of the goddess and then the man punishing her for eternity by reducing her existence to docile domesticity. She then in her defeat becomes a servile companion to her conqueror, the "upper-caste" man who is a staunch devotee of the goddess. The winning formula to tame and conquer the *Yakshi* was to offer her some lime smeared on the tip of a knife, which she always asks from her unsuspecting victim to chew her betel leaf. The symbolism of the knife, a phallic symbol with white lime at the tip suggests sexual liaison. There are also stories in the *Aithihyamala* in which the *Yakshi* falls in love with her victim; but she can never be part of his family rituals as a wife though she lives with him and bears his children. Some of the other stories also narrate *Yakshi* becoming *Devi* or goddess for going against her *swa-bhava* (her dominant traits) and attaining goddess like temperament and powers in benevolence, generosity and blessing. The stories imbue the *Yakshi* with particular powers to negotiate the good/evil binary and locate themselves in the narrative in-between space.

These narratives also unwittingly project the goddess and the *Yakshi* in similar terms. Both are portrayed as women of incomparable beauty. The former's beauty is a source of serene bliss and the latter's beauty entices to doom. Both are fond of adornments and blood sacrifices and have immense protective and destructive potentials. In almost all temples and *kaavus* of the goddess, the *Yakshi* is given a place in small shrines, outside the temples of the goddess. The goddess and the *Yakshi* are conflated to a singular entity worthy of worship in the *kaavus* of Kerala.

An interesting aspect of these stories in the *Aithihyamala* is that the Bhagavati and *Yakshi kaavus* were places of economic prosperity. The *kaavus* acquired wealth in terms of land, gold, and elephants. The economic functionality of acquiring wealth by these *kaavu* spaces narrated in these myths indicate that they were not just places of worship and confirm the dominant role of faith in deciding temporal matters. The goddess and the *Yakshi* would communicate through the oracles (*velichappadu* belonging to a specific community; only these community members became oracles) and certain events and happenings were interpreted as miracles. Diseases and famine were seen as punishments for the crimes and wrongs committed by society by the wrathful goddess. For redemption and exoneration of sins, the possessed oracles of the goddess demanded and got valuable gifts of gold, elephants, and land. The stories reiterate that without these offerings, the peeved goddess and the *Yakshi* refused to be pacified. They also sometimes demanded human and animal sacrifices. The power structures in the native societies were highly hierarchized and the tensions between the priestly and monarchical powers can be read as a subtext in many of these stories, especially with reference to land rights. The stories of the goddess and the *Yakshi* in the *Aithihyamala* give not only glimpses of the society, customs, and modalities of temple administration in precolonial Kerala but also about the social distribution of land and wealth and of caste and gender roles at that time. The *Aithihyamala* narratives have some unique features. Most of the *sthalapuranas* of the *kaavu* narratives conflate the particularly contextual, social, and cultural constructs of the *devi* and the *Yakshi*. The stories extoll the virtues of the Brahmin and other high caste males and their role in the maintenance and administration of the *kaavu*. Most of the *kaavu* stories end up giving second position to the *Yakshi* in the *kaavu*, though the kaavu may have been hers originally and the *devi* may have occupied it using her might. All *kaavu* stories emphasize the fact that kaavus were not just places of worship but also a space to become wealthy either using the benefaction of the deity or using the deity given boon to achieve money and fame by the extraordinary expertise of one representative member of the family owning the kaavu in learning, healing, prowess in art and architecture, cooking, astronomy and astrology, and so forth. The stories of the *Aithihyamala* captures a disjoint Zeitgeist and never mentions the focus of the *kavus* for which it is known for today as serving from time immemorial, a testimony to the ecological consciousness of the Malayalee. Historically, the sacred groves may have been places of worship of the people of ancient Kerala, during the Sangam Age, who followed "Dravidian practices which were not based on any religion. Their life was a mixture of primitive rites and rituals. They worshipped totem gods and spirits of the mountains, water, trees etc. They however, offered elaborate sacrifices to a particular goddess- the war goddess *Kottavai*" (Menon 2008, 81). These sites of *kottavai* sacrifice may be the origin of the *Yakshi* kaavu. When Buddhism reached Kerala in the third century BCE, the monks and nuns who were proficient in healing practices, Buddha himself was known as a healer, would have used, maintained, and preserved the Dravidian sacred groves as repositories of valuable medicinal plants and as Buddhist shrines. The Buddhists were Naga worshippers and therefore some of those groves could also have been refigured as *Sarpa Kaavu* by the monks and nuns. S. Menon explicates in his book, *A Survey of Kerala History* (2008, 78), that Buddhism impacted Kerala life and culture in deep and profound ways. He further elaborates in the same book on the onset of Aryanization of Kerala in 1,000 BCE and which reached a decisive stage by the fourth century BCE. According to Menon, Aryanization of Kerala was "the final submission of the local Dravidian races to the superior intelligence and administrative skill of the Brahmins of the north" (Menon 2008, 88). The first batch of Vedic Brahmins would have reached Kerala in the third century BCE itself. There are according to Menon evidences in Sangam literary texts vouching for Aryan influence in Kerala society especially when the noted Sangam poets were celebrated Brahmin saints patronized by Sangam rulers. In the eighth century AD, Aryanization of Kerala had reached its climax, with Brahmin scholars meeting the Buddhists in argument, completely defeating them and establishing the supremacy of the Vedic faith. They founded schools for the propagation of the Vedic faith and found a great champion in Adi Shankara and his works. Menon quotes Bishop Clad well to show the deep impact of Aryanization of early Kerala society and culture. He quotes, "The Aryan immigrants to the south appear to have been generally Brahminical priests and instructors, rather than Kshatriya soldiers, and the kings of the Pandyas, Cholas, Kalingas and other Dravidians appear to have been simply Dravidian chieftains, whom their Brahminical preceptors and spiritual directors dignified with Aryan titles, and taught to imitate and emulate the grandeur and the cultivated tastes of the Solar, Lunar and Agni-kula races of the kings" (Menon 2008, 90) The zealous Brahmins also foisted caste system on a casteless society and spread the Aryan ideology based on the primacy of *Chaturvarnya*. The Aryan missionary zeal in propagating its ideology found the Brahmins on a militant propaganda against Buddhism and Jainism. Consequently, these religions lost their following among the people. They also took to large-scale destruction of Buddhist *viharas* and shrines and established Hindu temples sometimes in the same place with a view to popularizing the Hindu religion. In the new Hindu temples, the worship of Hindu Gods like *Vishnu* and *Shiva* was made popular. Non-Aryan deities and practices were coopted into the Hindu fold and these temples. The Dravidian Goddess *kottavai* was accepted in the form of *Durga, Kali,* and *Devi Bhagavati* into the Hindu Pantheon. The reason why *Aithihyamala* kaavu narratives conflated the *Yakshi* and the *Devi* was in this subsuming act of the Dravidian *kottavai*, as the *Yakshi* into the Hindu Pantheon albeit in a secondary position. "The non-Aryan *Shasta* was represented as *Hari-Hara Putra* (Son of *Hari-Vishnu* and *Hara-Shiva*) in order to make this deity acceptable to both *Vaishnavite* and the *Saivitie* sections of the Hindu population" and to the new Hindu converts from Buddhism (Menon 2008, 92). A reason why the Sabarimala issue cannot be easily resolved rests in the Dravidian-Aryan conundrum associated with this erstwhile *Shasta kaavu*, which was appropriated as a Hindu temple. The Hindu religion in precolonial Kerala was forged by a synthesis of the Aryan Ideology of the north and the Dravidian ideology of the south.

The *Aithihyamala Yakshi-Devi kaavu* narratives point to a particular cultural history of assimilation and synthesis wrought about by hegemonic power dynamics similar to the Hindu nationalism in contemporary India. The *kaavus* of Kerala became private spaces maintained by upper caste Hindu families, which were at odds with the original Dravidian understanding of the space as a casteless, preserved, communal place of worship without economic functionalities attached to it.

THE *KAAVUS* IN THE COLONIAL TIMES

In the article "Social and ethical Dimensions of Environmental Conservation," M.D. Subash Chandran and T. V. Ramachandra write, The British colonization of India led to large-scale systematic exploitation of forest resources, which had a tumultuous effect on the conservation ethics of the local people. Most of the extensive forests of the country, teeming with wildlife, were taken over by the government during early nineteenth century and exploited commercially. Cleghorn (1861), the first British Conservator of Forests in South India, stated the government takeover of forests was "somewhat ill-advised attempt . . . thoroughly failed in its objective . . . and threatened the speedy and complete destruction of forests themselves." The reserved forests were meant to meet the needs of the urban, industrial, and military sectors, and the protected forests those of the rural population (Chandran and Ramachandra 2008, 7). This approach toward natural resources by the Western colonizers had its origin in the Christian faith they followed. In the Western mind, the prevalent view of nature was shaped by the Bible and some biblical injunctions. They believed that God created nature for human's use and benefit. In the Book of Genesis, God instructs Adam and Eve to be "fruitful and multiply and fill the Earth and subdue it; have dominion over everything that moves upon the Earth." This biblical injunction supported a dominant tenet of Western philosophy: nature should be converted into wealth as quickly as possible and used for the benefit of the people. This perception justified nearly all of their land uses and reinforced the belief that to leave land unused was to misuse God's gift and a sinful mistake. A popular belief in medieval

Europe was that the wilderness or unused land was inhabited by evil spirits or monsters, in contrast to the orderly appearance of agricultural landscapes. In practice, the wealth and benefits that came from this policy from colonies helped the citizens of the colonial powers to accrue more, while the needs of the natives were largely disregarded. The contradictory worldviews regarding nature and its resources utilization had far-reaching impact on the identity and existence of the *kaavus* in Kerala. In the colonial classifications and land settlement records, property and land appear either in relation to forest or in relation to revenue or agrarian/taxable territories. For the colonial administrators, all understanding of property or land was restricted to these realms; other spaces were classified either as religious or social. The religious or social were implicated in economy, however not in administering them as groups, but transactions, exchanges, and understanding of value were strictly confined to economy. The *kaavu* was strictly tied to the realm of religion by the British administrators and. classified as noneconomic by colonial classifications; however, property, value, and territory were being reproduced through these sacred spaces as well.

There were events in colonial records that illustrated the resistance and reluctance of tribal communities to violate the property of their deities in the kaavus, when they were asked to fell trees there and this reluctance was seen as obvious signs of "animism" and "barbarianism." Behind the British concept of nature was the idea of temporal progression and this progression was seen as facilitated through the passing of seasons. Nature was divided into predictable temporal compartments, that is, the four seasons. This predictable and regulated movement of nature progressed into days, nights, and years thus indicating a natural and linear movement of time. This idea of temporal progression was at the core of certain colonial attitudes toward nature. The ability to appreciate nature was seen as a sign of progress; the capacity to adore nature was a feature expressing the fineness of minds. But worshipping nature and superimposing human characteristics to nature (and supernatural characteristics to humans) were perceived as an obvious sign of lacking this temporality (Moorkoth 2012, 16). The British administrators believed that sacred groves are widespread phenomena among the tribal groups of India and a common feature of these groves was that they are held sacred. There were two ways in which they understood this phenomenon. Firstly, the sacred groves stood for the religious sentiments of the tribal groups. Sacred groves dedicated to local deities, or ancestral spirits, are protected by local communities. This view developed from the colonial land settlement officers who didn't want to interfere with the religious activities of the natives.

Records from colonial Malabar show instances of how the colonial governments were in favor of protecting such lands within the forest. One interesting example is a case from Malabar where the Forest Department officials were digging a well but met with no success in finding water. It was suggested that the solution to this problem is to do a pooja (worship) for the local deity in the nearest grove. The chief conservator of the forest issued immediate order to do the pooja so that the problem can be tackled. The second view is that sacred groves stand for biodiversity and the sustainable practices of tribal groups. This view emerged primarily due to the interest in ecological concerns in the post-independence period. With the failure of the colonial developmental models and with the rise of the biotechnology industry worldwide, local sustainability systems were revisited. As a result, the adivasi groups and their sustainability models acquired a special place in the new projects modeled on local development (Moorkoth 2012, 24). The colonial administrators respected the sacrality of the *kaavu*, but they never stopped the local landlords owning the groves from exploiting the natural resources for commercial purposes. The *kaavus* were seen as cultural constructions depending heavily on local consensus. In Kerala, while some *kaavus* are dedicated to deities, others dedicated to snakes and some others were dedicated to ancestors and ghosts. They varied in size, function, practices, customs, and taboos. And so, rather than any ecological or religious precept, what determined the functionality and the structure of kaavu in the colonial times was the local power relation and social dynamics.

The colonial experience was central to the origin of the contemporary conservation movement in India. Colonial knowledge in combination with imperialist expansion was instrumental, in the establishment of protected areas in India. Across the tropics, initial colonial conservation agendas were rarely overt in their economic and political intent; game and forest laws were articulated as being necessary to overcome the inefficiency of native people (projected variously as rootless, ignorant, recalcitrant, and savage) to manage their own resources (MacKenzie 1988; Wolfe 2006; Sridhar and Oommen 2014).

THE POSITION OF THE POSTCOLONIAL *KAAVUS*

Many original *kaavu* owners like the untouchables and tribal have been driven out from their lineage groves in the plains and forests in the hills. These places have now become sites of inquiry and settlement for encroachers from the plains; students of environmental science; nongovernmental organizations (NGOs); Hindutva activists and tourists with different causes, purposes, and reasoning who now have access to the forested mountains and groves. A commonality that binds them however is their "cause of ecologisation of nature, and a related moment of its commodification with a nationalist twist of Hinduising non-Hindu deities of the hills" (Uchiyamada 2001, 131). Thus, the deregulation of Indian economy, through the central government intervention, is taking place side by side with the regulation of minor deities along the meta-narrative of the Vedic order of things. There are regular readings organized of *Ramayana* and *Mahabharata*, the prime politico-poetic narratives of pan-Indian Hinduism, reverberating from loudspeakers tied to *aatmaavu* ("soul") trees in various ancestral shrines and *kaavus* of tribals in the plains and the hills. This repetitive and disciplinary performance of offering *Ramayana* is at the heart of the Hinduization of Malayalee landscape, and is part of a related process of the making of the postcolonial Hindu national state inscribed in the latter (Uchiyamada 2001, 131). As Gilles Tarabout succinctly put it, "when ancestral 'sacred grove' are either cut down or transformed into Hinduised shrines and disappearing from the life world of Malayalees, the imagined 'Indian sacred groves' are emerging in the discourses of national ecological history, and of religion and ecology." Ecotourism is developing, attracting town people to forests in the mountain where Malayali trekkers and members of environmentalist groups will be "camping in the hills, studying about environment, meeting with hill tribals, enjoying birdwatching." Indeed, the Kerala Tourism Department promotes online an ecological image of Kerala which relies not only on the Reserve forests of the Western Ghats, but on "sacred groves" as well: "When the land, religion, myth, culture and civilization harmoniously blend together in a small space replete with greenery, we call it kavu a unique and ancient ecological haven common to the land of Kerala" (Tarabout 2015, 25).

CONCLUSION

Claims about antique environmental awareness embedded in Hinduism, holding up the iconic sacred groves as evidence, are an effective rhetorical device. Actually, however, Brahmanism had configured the *kaavus* in Kerala in such a way that its value and power were best harnessed when limited to tiny spots. The groves were controlled by land-rights owners. He was for all purposes, supposed to be the patron and protector of such sacred spaces. However, in practice, once the political power associated with these offices was perceived, and the higher cash values of a new land-market were introduced, scant regard was paid to the religious sanctity of these sites (Freeman 1999, 291). Cultural nationalism propagates a pan Hindu identity, which homogenizes religious Hinduism overlooking variegated caste and geographical differences understanding and practice, tricking the lower caste and tribal citizens into embracing Hinduism and by making Hinduism a money-making scheme. In the case of the *kaavus*, the owners have understood the economic plausibility of converting them into money spinning schemes. This also helps fringe elements associated with the groves financially. The first step in Hinduising a grove is to fell the trees and clear the spot for temple constructions. "For priests of various castes, astrologers, and small-scale religious entrepreneurs, clearing sacred groves is booming business" (Notermans et al. 2016, 14). In this business model of grove entrepreneurship, there is no relationship lost between the clearing of groves and a loss of religion. Instead, the clearing "resulted in an intensification of both religion and the grove's economy, with more and more temples built, rituals done, offerings given, and the demands of the gods becoming ever more complicated and compelling. The grove's wealth developed in 'a sacred economy' " (Notermans et al. 2016, 14). Thus, the postcolonial Kerala *kaavu* is located in the interstices of religion, economy, cultural nationalism, and natural resources management, making it a liminal space.

BIBLIOGRAPHY

Athreya A. 2016. "Cultural Nationalism in India." *Anthropol* 4: 165. doi: 10.4172/2332-0915.1000165.

Chandran, M. D. S. 2016. "From the Shadows of Legitimacy Problems and Prospects of Folk Healing in India." *Journal of Traditional and Folk Practices* 2 (3–4): 74–95.

Chandran, M. D. S., and J. D. Hughes. 1997. "The Sacred Groves of South India: Ecology, Traditional Communities and Religious Change." *Social Compass* 44 (3): 413–427.

Chandran, M. D. S., T. V. Ramachandra, and Environmental Information System, Sahyadri (ENVIS). 2010. "Social and Ethical Dimensions of Environmental Conservation." *Sahyadri E-News*: 2–31.

Freeman, J. Richardson. 1999. "Gods, Groves and the Culture of Nature in Kerala." *Modern Asian Studies* 33 (2): 257–302.

Gadgil, M., and V. D. Vartak. 1975. "Sacred Groves of India: A Plea for Continued Conservation." *Journal of the Bombay Natural History Society* 72: 314–321.

Kumar, Sudarsana. 2018. "Substantive Economy of the Hill Tribes of Kollam & the Indigenous Trade Network in Early Medieval Kerala (800–1500CE): An Appraisal." *Journal of Indian History and Culture* 24 (64–97): 65.

Menon, A. Sreedhara. 2007. *A Survey of Kerala History*. Kottayam: DC Books.

Moorkoth, Jensy Meera. 2013. *Countering Hegemonic History: Understanding Adivasi Land Politics in Kerala*. 1st Edition. Ciudad Autónoma de Buenos Aires: CLACSO. Notermans, Catrien, Nugteren Albertina, and Sunny Suma. 2016. "The Changing Landscape of Sacred Groves in Kerala (India): A Critical View on the Role of Religion in Nature Conservation." *Religions* 7: 38. doi: 10.3390/rel7040038.2-14.

Ramachandra, T. V., M. D. S. Chandran, K. V. Gururaja, and Srekantha. 2006. *Cumulative Environmental Impact Assessment*. New York: Nova Science Publishers, Inc.

Sankunni, Kottarathil. 1974. *Aithihyamala*. Volumes I & II. Kottayam: Kottarathil Sankunni Memorial Committee.

Shekar, Ajay. 2020. "Panayannar Kavu: The Sacred Grove by the Pampa Near Parumala and Niranam." https://ajaysekher.net/tag/buddhism-in-kerala/html. Accessed 27 March 2020.

Sivaramakrishnan, K. 2020. "Nationalisms and the Writing of Environmental Histories." www.india-seminar.com/2003/522/522%20k.%20sivaramakrishnan.html. Accessed 26 March 2020.

Sridhar, A., and M. A. Oommen. 2014. *Representing Knowledge: LEK and Natural Resource Governance in India*. Deutsche Gesellschaftfür Internationale Zusammenarbeit, German Federal Ministry for Environment, Nature Conservation, Building and Nuclear Safety and Dakshin Foundation, 130 p.

Tarabout, Gilles. 2015. "Spots of Wilderness: 'Nature' in the Hindu Temples of Kerala." In *Rivistadegli Studi Orientali*, Fabrizio Serra editore, *The Human Person and Nature in Classical and Modern India*, edited by R. Torella and G. Milanetti, Supplementoalla Rivistadegli Studi Orientali, n.s., vol. LXXXVIII, pp. 23–43.

Uchiyamada, Yasushi. 2001. "Journeys to Watersheds: Ecology, Nation and Shifting Balance of Malayalam." *Journal of the Japanese Association for South Asian Studies* 13: 107–141.

Chapter 4

Politics, Oil, and Theater in Africa

Stephen Ogheneruro Okpadah

> Africa is certainly the hot spot of the resource-conflict link: "Blood diamonds" in Sierra Leone, coltan and diamonds in the Democratic Republic of the Congo (DRC), oil in the Niger Delta in Nigeria and oil and diamonds in Angola.
>
> —Basedau and Wegenast (2009)

The African continent is rich in natural and human resources. This accounts for her encroachment and colonization by the West. This transcontinental and transcultural encounter led to the upsurge of the trans-Atlantic slave trade, which the continent encountered for more than three centuries. Kingdoms in present day Nigeria, Tanzania, Ghana (formerly Gold Coast), Cote D' Ivoire (formerly Ivory Coast) among others suffered the brunt of this psychological and demeaning experiences created by superpowers such as Portugal, France, and Britain. During the slave trade, "The humanity and self-esteem of the slaves were reduced to a bare minimum if not completely eliminated, for instance, at the time they were captured and during their transportation" (Cornelli 2012). These traumatic experiences are well documented in the literary piece of one of the first educated slaves from Africa, Olaudah Equiano, in his book, *The Interesting Narrative of the Life of Olaudah Equiano, or, Gustavus Vasa, the African*, published in 1789.

The abolishment of the slave trade was the genesis of European colonization of Africa. However, the final blow came with the Berlin conference. McGuffe (2016) notes that the "identifying boundaries between inside and outside Africa are manufactured for imperial interests—just as the boundaries of African nations were arbitrarily determined at the Berlin Conference in 1885." This conference that regulated "European colonization and trade in Africa" (Cornelli 2012) was an attempt to subdue Africa and her natural resources. We must recall that the colonial masters turned toward Africa to source for the raw materials for their industries and the development of Europe. In South Africa especially during the apartheid regime, the colonialists' exploitation of human and nonhuman resources came to play. The indigent South Africans were made to extract the gold in the mines. Walter Rodney's seminal research, *How Europe Underdeveloped Africa*, fully captures this exploitative enterprise.

The decolonization of Africa after World War II changed the political dimensions of natural resources and economic processes in Africa. It was explicit that independence was not economic inclusive. The natural resources were still controlled by the imperialists. In Nigeria, for instance, the natural crude oil was discovered in Oloibiri, a present settlement in Bayelsa state in 1956. British Shell Petroleum saddled with exploring and exporting the product still continued with this responsibility after the October 1, 1960, liberation exercise. The discovery of oil in other African countries such as Algeria, Libya, Egypt, Sudan, Angola, Republic of Congo, and Ivory Coast, came with Western and American interference. The romance between the United States and Saudi Arabia, motivated by the Middle Eastern country's oil, was also replicated in these African countries. Unfortunately, any attempt to resist the hegemony of America and the West over oil-producing countries in Africa received stiff resistance and stringent penalty. Suffice to state at this juncture that the Libya revolution and the decades of war in Sudan were wars for oil, and the role of America and the West in these wars are explicit. This was the fulcrum for the politics of oil in Africa. Hence, oil has become a prime mover of global politics.

For more than five decades, the African literary enterprise has captured the intersection of colonialism, politics, and natural resources. Most of Ngugi Wa Thiong'o, Peter Abrahams, Ebrahim Hussein, and Alex La Guma's proses and plays were an affront on the exploitation of natural resources in Africa by the Whiteman, even after the purported independence of the British and French colonies. The imperialist's politicization of natural resources is evidenced in the colonial master's confiscation of lands from natives of Jomo Kenyatta's Kenya and Robert Mugabe's Zimbabwe. Postmodern African theater artists have expounded on the colonization of natural resources by the West. However, the approach of dramatists, such as Ahmed Yerima and Sam Ukala, among others, is multidimensional. Their plays are critical of European and American interest on natural resources in Africa—especially oil, and also navigate the politicization of oil in Africa by the African bourgeoisie. The utility of drama and theater in locating the intersection of politics and oil in Africa cannot be overemphasized as Gordons Vallins (1971) attests to this in his assertion:

Not merely can drama reflect and reinforce our attitude in values, it can also attempt to change them, to shape the culture in which it exists: it can be used to suggest alternatives to the present systems, it can give warnings, it can explore the relationship of people subjected to the complexities of the system and thus expose its prejudices and injustice.

The advent of independence in Africa coincides with the mass discovery of oil in countries, such as Nigeria, Sudan and other African nations. This discovery with its complexities brought about the popularity of the oil factor in postmodern African drama. With the capacity of drama and theater to influence it viewers, it became imperative to use these media (drama and theater) to explore the oil question, an issue that is prevalent in the society. To this end, the questions that emanate in this chapter are, to what extent does oil connect Africa and to its European and American counterparts? What is the interface of oil and crisis in Africa? How have African dramatists captured the politicization of oil on the African continent? While this chapter seeks to find the answer to the earlier raised questions, this study also examines the interface of oil and crisis in Africa. It investigates how African dramatists have been able to present the political dimensions of oil exploration and crisis in Africa.

POLITICS AND OIL IN AFRICA: AN INTERFACE

Africa is the home of numerous countries with crude oil. Nations such as Angola, Nigeria, Libya, Sudan, Algeria, and Republic of Congo thrive on oil. Foreign policies of oil-producing countries in Africa are also built on the liquid black gold. Politics and oil in African countries are interwoven. Oil has further integrated Africa into the America, Russia, and China divide and battle for supremacy. America has long been at the forefront of oil politics. His post–World War II pact with Saudi Arabia for the exploration of the latter's oil in return for protection from her (Saudi Arabia) enemies fully captures the former's attempt at Americanizing the world oil enterprise. America's race for acquisition of African oil is not a recent phenomenon. China is also well known for its petrol-colonial tendencies. "In 2005, China imported nearly 701,000 bpd (barrels per day) of oil from Africa, which was approximately 30 percent of her total oil imports" (Billion 2013). The exploitation of Africa's resources has been the norm from precolonial times. Recall Britain's looting of the art treasury of the Benin kingdom in the wake of the British-Benin war in 1897. Other Kingdoms in Africa also suffered this fate. The land acquisition methods in Zimbabwe and Kenya were also part of the colonial invasion of Africa and her resources. The "transnationalization of oil has created spaces for the emergence and growth of oil war as expatiates that oil has created a fluid, dark foundation upon which many of the world's wars, conflicts, and grievances have been based" (Gonzalez 2010). Inter-states conflicts (the Nigeria-Cameroon battle for Bakassi Pennisula) and intra-state war such as in Sudan were oil motivated. These conflicts are part of the Euro-America exploitation engagement. What comes to the fore at this juncture is the dominant role the United States and other world powers play in these conflicts. The West has not ceased to be powerbrokers in the oil conflict in Africa.

The role of America in the Libya's uprising, which led to the collapse of Muammar Gaddafi's regime, America's stand on the decades long Sudan civil war and Egypt's revolution reveals that the West has always been the prime mover of oil war in Africa. Numerous reasons account for the politicization of oil by America and the West. Oil is an important commodity, which lubricates nuclear industry and advances capitalism. The transportation sector also thrives on oil. In fact, "in the next quarter century, the number of vehicles worldwide is projected to rise from 700 million to 1.3 billion; twenty percent of that increase in China alone" (Victor 2004). "We remain highly dependent upon oil and are likely to remain so for the next several decades. Oil provides about 40 percent of global energy needs and is projected to provide about that amount in 2030" (Adam 2005). The global consumption of oil has risen considerably. By "2025, oil consumption is projected to be 119 million barrels per day, up from 77 million barrels in 2001" (Yergin 2006). Michael Clare, a professor at Hampshire College, articulates that America prioritizes oil. In 1945, the King of Saudi Arabia Abdulazeez Ibn Saud demanded protection for his country from president Roosevelt of America, in return for oil. America also showered the king and his 3,000 princesses with gold. What comes to bear in Professor Clare's statement is that the Western nations and America do not only struggle to explore oil, they also put measures in place with which oil can be got at a cheap or not cost. Oil companies sprang up to take charge of its exploration. These oil companies were owned by America, Britain, and France, among other super powers. This second phase of colonialism after flag independence depleted the lands explored and unfit for human health and at the same time appropriates the capitalist ideology. In other words, oil played a major role in stabilizing capitalism as against African economic processes such as Julius Nyerere's *Ujaama*—an Africa model of scientific socialist aesthetics.

The huge sums of monies acquired from oil exploration, refining, and distribution have only brought the billionaires conspiracy to the fore. This conspiracy is the continual depletion of the environment at the expense of the poor occupants of territories in which oil is explored. The multinational oil companies are also in a tussle of supremacy. Paradigmatically, in Tehran, in the early twentieth century, the oil companies fought bitterly here to control oil. The first oil companies, known as the Seven Sisters—*Exxon, Mobil, Chevron, Texaco, Gulf, Royal Dutch Shell*, and *British Petroleum*—have been in the struggle for the allocation of oil rich lands. The Seven Sisters, most of which were nations owned became more powerful than the indigent African countries whose oil was being explored. Pierre Terzian, Director, Petrostrategies in Saudi Arabia, in an interview in the documentary, *The Secret of the Seven Sisters*, makes us understand that these people bought up whole countries or at least their resources with documents barely fifteen or twenty pages long. The Seven Sisters cartel determined the global price of oil and distribution of the product. In 1962, the cartel was summoned to face the Supreme Court for the first time. But the investigation was blocked and the trial never took place. What is good for the Seven Sisters is good for America. The power of the cartel was visible that the Sheikh Ahmad Zaki Yamani, Saudi Arabia minister of petroleum (1962–1986), affirms that they were a state within a state. These oil companies produce and export the oil the price they want. The countries where oil was being explored had no power or say in the oil business.

Petrol-imperialism in Africa is multidimensional. While exo-colonial engagement remains obvious, endo-colonialism is more destructive. Exo-colonialism comes to play with the activities of the transnational oil companies. These Western-owned companies have dislocated the indigent inhabitants of oil-owned communities. This is explicit in the Sudan situation. The "impact of the discovery of oil in southern Sudan has been to create demand for the lands of the pastoral communities living there and practicing their traditional livelihoods of cattle rearing and agriculture. As a result, the oil fields are the heart of the contested area between the forces of North and South" (Switzer 2002). Indigenes are made to relocate from oil field, or remain there to the detriment of their health. The tussle in the oil rich Niger Delta region of Nigeria is not farfetched. The Urhobo, Isoko, Ijaw, Igbo, Itsekiri, Ibibio, Efik, Ogoni, and Ekwerre people during the pre-oil times were known to be commercial farmers, hunters, and fishermen. Unfortunately, the discovery of oil and subsequent encroachment of transnational oil companies led to the dislocation of their means of livelihood. Perhaps, this accounts for the high rate of poverty in the region. "The discovery of oil has also led to deforestation and ecological degradation, threatening the renewable natural resources and the ecosystem services in a number of ways" (Babatunde 2010). The environment remains degraded. The World Wildlife Fund (2005) notes:

> The World Wildlife Fund considers that it is one of the most polluted places on the face of the earth. Nigeria has the highest gas flaring rates in the world; until recently over 80 percent was flared as a result of drilling. A majority burn for

24 hours a day with some having remained active for 40 years. This has a direct impact on the local ecology, climate and people's health and property.

The state of the environment and the poor economic situation has led to an upsurge in migration from the region. But it is distasteful to note that indigenes of the region that produces Nigeria's major source of income migrate to Northern and Western urban settlements, such as Lagos, Kano, Kaduna, and Abuja, for greener pastures. These towns (except Lagos) largely depend on the proceeds of the oil from the Niger Delta for survival. Apart from Portharcourt in Rivers State and Warri in Delta state (both in the Niger Delta region), no other geographical space in the Niger Delta region can boast of socioeconomic and infrastructural development. Suffice to state that in the past one decade, Warri's economy has been a shadow of itself while Portharcourt has become a hub of delinquency, cultism, and prostitution.

A second dimension to the oil miasma is the endocolonial engagement of the region by its elites who stand as intermediary between the region and transnational oil companies/government. Endocolonialism in the context of this study is a process whereby an individual, a group of people, or a central system of power within a society subdues human and nonhuman resources of the society. This implies that the endocolonizer is a part/an indigene of the endocolonized. The endocolonizer is the binary of the exocolonizer who is a guest of the colonized. Monies and resources provided by the government for the development of the region have often times been siphoned by the elites for their own gains. The multinational oil workers are well paid and their squandermania practice has only promoted the capitalist culture in the region. Although undeveloped, inflation rate of the region remains high.

In Nigeria, "oil has created a clash between the local people on the one hand and the national governments and multi-national oil corporations on the other over the rights to oil revenue" (Gonzalez 2010). The government's response to this growing anger among the youths in the "marginalized region has been both brutal and violent, and this has further fueled the hostility within the Delta" (Bromwen 1999). The imprisonment of Isaac Boro, the execution of Ken Saro-Wiwa and the Odi Massacre carried out by the Nigerian military on November 20, 1999, are few examples of the reaction of the Nigerian government to the agitation in the region.

In Sudan, although social and political factor were instrumental to the war, oil played a major role in the crisis that raged for decades. Lundin Oil, a Swedish oil company, has a 40 percent stake in Block 5A. "Allegedly, in order to guarantee the safety of the oil company's operations and clear area for a road to the concession, the government waged war against the local communities, whom were forcibly evicted and their villages razed" (The Editor 2001). On the other hand, while Northern Sudan had a major say in the governance process, the oil-rich South was subdued by their counterpart. The North's refusal to the South's separatist attempt was a result of the large oil deposit in the South. This is a prime reason for Khartoum's resistance to southern independence. Additionally, oil is considered the single leading factor in the perception of the Sudanese civil war. The ownership of these oil reserves "were controlled by Khartoum and it has been the Sudanese regime's refusal to give the south a say in the development of the fields that was one of the original causes of Sudan's 16-year civil war" (Sudan Oil 2001). Recall the Twelve-Day Revolution of Isaac Boro in which his militant group declared Niger Delta Republic. This was the result of the region's neglect by the government and the oil companies. This portends that oil has been a played a major role in Nigeria and Sudan's separatist politics.

It is unfortunate that oil, of which revenue can be utilized to better the lives of the majority of citizens in these three countries, has only led to suffering, strife, ruin, and war, and it is the innocent that has and will continue to suffer the most. The complexities, failures, and successes in politicizing and depoliticizing oil in Africa have been replicated on the African screen and live theaters. Consequently, while African theaters on the oil motif are examined in the next part of this study, particular reference is made to the Nigerian theater. Furthermore, in this study, the terms dramaturgy, theater and performances are used interchangeably.

POLITICS, OIL, AND THEATER IN AFRICA

African literary artists have engaged in discourses on politics and the oil rift. This has been expressed in postcolonial literary texts. The revolutionary spirit of postcolonial African texts was the springboard for a theater of radical praxis for the oil question. We should recall that before now, the works of Ousmane Sembene (*God Bits of Woods, Ceddo, Xala*, etc.), Ebrahim Hussein (*Kinjeketile*), and Ngugi Wa Thiong'o (*Weep Not Child* and *The Trials of Dedan Kimathi*) among others had the revolutionary motif. They were attempts to fight back at the imperialists, their policies, and hegemony over their colonies. These literatures were anchored on decolonizing all facets of the colonized societies as Wa Thiong'o appropriated as a "decolonization of the mind." These creative works were located in the context of liberatory aesthetics.

The 1970s that saw the rise in independence of numerous British and French colonies in Africa gave rise to a new generation of literary artists whose goal was a critique of the failure of postcolonial African leaders whose governments were marred with corruption and maladministration. Although Ayi Kwei Armah's *The Beautiful Ones Are Not Yet Born* and Chinua

Achebe's *A Man of the People* reveals this trend, it is more apt in the plays of Femi Osofisan, Kole Omotoso, and Bode Sowande. These plays were product of Karl Marx's materialist divide. The radical praxis of the above writers led to an easy transition into the dramaturgy of oil crisis in Africa. Foremost among literary artists whose works focus on politics and the oil crisis are Helon Habila, Ogaga Ifowodo, Ibiwari Ikiriko, Nnimmo Bassey, Emmanuel Dandaura, Tanure Ojaide, Ahmed Yerima, and others. All but Habila and Dandaura are indigenes of the Niger Delta. Hence, a firsthand knowledge of the oil situation may have helped in creating factual narratives that capture the Niger Delta condition. The works of these literatis, such as Yerima (*Hard Ground*), Ikiriko (*Oily Tears of the Delta*), Bassey (*We Thought It Was Oil but It Was Blood*), Ifowodo (*The Oil Lamp*), Stephen Okpadah (*The Holocaust*), Habila (*Oil on Water*), and Ojaide (*Delta Blues*), investigate the oil crisis and the exploitation of Africa by Africa and the West. The exploitation of Africa by Africa seems to be more explicit in these narratives.

The drama/theater enterprise has not been left out of the oil question. Playwrights especially from the Delta region have explored the oil crisis from various perspectives. J. P. Clark in *Wives Revolt* and Tunde Fatunde's *No More Oil Boom* attest to endocolonialism characterized with the new oil culture. In *Wives Revolt* Clark exposes the masculine bias attached to the sharing formula of the oil wealth in the Niger Delta. He reveals that while the women suffer more from the activities of the oil companies, the male divide the monies meant for the development of the region among themselves and place the women at the margin. However, Clark seems to present the marginalized women in his plays as a metaphor for the peasants in the Niger Delta who suffer the brunt of oil exploration, and the male gender as the elites who appropriate the monies meant for the development of the region. *Wives Revolt* is a drama of revolt, which advocates for the radical approach in achieving the desired liberation of the people of the Delta from oppression and dominance.

Simon Ambakederemo is considered the first Nigerian playwright to incorporate the oil motif in his work. According to Ben Binebai, "in Ambakederemo's *Isaac Boro* . . . the playwright and the play text raise and present the voice of the Marginalized and oppressed Niger Delta people" (Binebai 2015). Ambakederemo's play, *Isaac Boro*, published in 1978, is an account of the life and death of Isaac Jasper Adaka Boro of the Niger Delta extract. The play explicates identity politics in the oil war in the Delta. It investigates exo and endo-colonial processes. While the eponymous tragic hero, Isaac Boro criticizes the West for her contribution to the despoliation and suffering of the Niger Delta people and environment; he also reiterates that the elites who stand as intermediary between the government and Delta are sabotaging the war against ecological destruction and the development of the region. Isaac Boro, the eco-warrior is also well represented by Ben

Binebai in his play, *Drums of the Delta*, Sam Ukala in *Fumes of Fuel* and Yerima in his trilogy, *Hard Ground, Little Drops* . . ., and *Ipomu* as well as Akpos Adesi in *Agadagba Warriors*. All of these plays interrogate the petro-culture, environmental despoliation of the region by the transnational oil companies, eco-terrorism, government's response to eco-terrorism, and so on. *Hard Ground* focuses on godfatherism as a major player in the Niger Delta conflict. In this play, Nimi, the protagonist, fights for the emancipation of the region from the grips of the enemy—the oil companies and the military. Nimi and his boys are being sponsored by the Don—a metaphor for the Niger Delta elites. However, betrayal sets in as the Don's greed and selfishness leads to the death of Nimi's wife and child. The Don proposes a meeting with Nimi who is poised toward avenging the death of his boys. At the meeting, he successfully kills the Don who happens to be his (Nimi's) father, Baba.

In *Little Drops*, Yerima investigates the Niger Delta imbroglio from a gender perspective. The place of women in the struggle for the resolution of the Niger Delta conflict is examined. Yerima uses the Lysistrata paradigm—in the context of Aristophanes's play *Lysistrata*—to enlighten the womenfolk in the Delta region about their place in the struggle. *Little Drops* . . . a female-cast-dominated narrative also tells its readers-cum-audience that for the eco-war (militancy) to be fully managed in the region, the elites in the Delta, government, and transnational oil companies operating in the region must include the women in the conflict resolution process. The female characters in play, thus, reveal how they suffer the brunt of the warfare engaged in the creeks by the Police Joint Task Force and militants. They complain of the manner in which they are raped and slaughtered by the soldiers and the militants. Memekize, Queen Azue, Mukume, and Bonuwo in their complaint realize that to put an end to the politicization of the crisis in their home is their collective responsibility. They must not sit and watch the militants and government heat up the Niger Delta polity. The politicization of the oil war is also depicted in the character of king, Queen Azue's husband. The king is corrupt and as such, instrumental to the unresolved crisis in the region. He heats up the polity to realize financial gains from the government and multinational oil companies. The monarch's in the region promise the multinational oil companies and government that they would advise the militants to desist from attacking the companies and their properties. These promises are usually accompanied with huge gifts from the companies and government. Toward the end of *Little Drops* . . ., Kuru, a militant, tells Azue, the queen, that her husband exploited the region and sold the conscience and sense of duty of his people for money. In the Niger Delta, while the militants stay in the creeks in untold hardship, the monarchs and chiefs revel in exotic cars, luxurious buildings, and vacation to Europe, America, and Dubai, with the monies given to them to develop the region. Due to their personal gains, these actors do not want an end to the crisis in the region. In fact, they sometimes join in incarcerating the militants to enable the latter launch reprisal attacks on the government and the oil companies. This would enable the crisis linger on. The longer the crisis, the more money they make. Fortunately, the crisis is curbed with the government's call for amnesty and the militants' embrace of this process. This is fully explicated in the last play of the trilogy, *Ipomu*. The militants embrace amnesty and peace is restored to the region. However, the government and the oil companies do not stick to their end of the bargain.

In *Ipomu*, Ebiere, the queen, tells the prince, her son, Ipomu, that she is happy that he had returned home safely. Returning home in this context means discarding the life in the creeks, turning a new life from kidnapping of oil expatriates, indulging in crude oil bunkering, destruction of oil pipeline, killing of the soldiers and other security operatives, and many other vices. The play focuses on the amnesty program initiated by the federal government and how genuine it is. The agitation of the Niger Delta militants in *Hard Ground* and the advocacy for peace by the women in *Little Drops . . .* watered the ground for the actions in *Ipomu*. Thus, Ebiere sends for Ipomu who is in the middle of a game in the city. She tells her that he must not return to the creeks, else he will not return alive. She had seen him on television; that day he laid down his arms in the company of his cousin, Igege. According to Ebiere:

Ebiere: In the talking box in Yenogoa, the day they gave up their guns, and bombs, I did not recognize you at first. Then you took of your dark glasses . . . your image stamped to my brain, haunted me (124).

This is the image of the boys fighting for emancipation of the Delta from oppression by internal colonialists (Nigerian federal government and selfish chiefs in the region) and expatriates oil companies. However, the amnesty program had begun the development of some places in the region. On his return from Port Harcourt, Ebiere asks him about the situation in the Garden City. He submits that the city is now a changed place. Buildings everywhere. Bridges on the roads, new houses, and cars. In fact, he exaggerates that almost everyone has a car now. Oil money has brought traffic jam everywhere. While the queen assumes that the fighting effort of the militants is not futile, so many issues still remained to be attended to. There still remains a paradox. This stems from the fact that while those who fought for the liberation of the region are not given all the monies they were promised, other youths in the region, who were law abiding, are not integrated into the amnesty process. Does this portend that being law abiding has its negative implications? Do the government and the oil expatriates only understand the language of violence?

Undoubtedly, the federal government pays particular sums of monies to the oil-producing states. However, these monies pass through the fingers of a few. Corporate social responsibilities of the oil companies include the award of scholarships (from high school through undergraduate to doctoral level) to indigenes of oil-producing communities, establishment of hospitals, construction of good roads, among others. The leaders—upper class—are usually the beneficiaries of such. Oyeso contends that it would be better "only if more money could leave the hands of the individuals, and be used to develop . . . the villages of the inner creeks more." Focus is laid on already-developed places, such as Port Harcourt, Yenagoa, Asaba, Calabar, Uyo, and Umuahia, among other places, at the expense of the interior village where the oil is explored. We submit that to achieve the aim and objectives of the amnesty program, the government must strike a balance in the blue print of development in the region. The notion of developing urban areas and placing the rural areas in the margin should be deemphasized. Akpos Adesi's *Agadagba Warriors* is another play in this category. This play reveals "the mismanagement and incongruous sharing formula of the country's resources between the land owners and the supposed lords" (Julius-Adeoye 2013).

The Nigerian government has engaged in numerous exercises in an attempt to clean up the Ogoni environment. "Ogoni land which sits between Port Harcourt, the oil capital of Nigeria and home to Shell Nigeria and Bonny Island where the main oil-export terminals are located" (Akinboka 2011), is environmentally degraded from oil exploration by Shell British Petroleum and other oil companies. This has been the norm for more than three decades. This norm has not only dislocated the people who are predominantly farmers and fishermen, to compound their misery, the government have silenced their attempt to speak out with the platform of MOSOP headed by Ken Saro-Wiwa. The latter's activism against the despoliation of the environment of the Delta led to his martyrdom in November 1995. Environmental degradation, a precedent of the Ogoni crisis, still remains prevalent in the Niger Delta region. The setting of Yerima's trilogy is the Nigerian Niger Delta. Nevertheless, its preoccupation reflects what holds sway in all oil-producing countries in Africa. Community theater projects have also been carried out by theater artists especially in the academia with the aim to educating and informing peasants about the need to desist from pipeline vandalism as well as to serve as a mediator to the oil crisis where other media has failed.

CONCLUSION

The West has been the prime mover of the oil crisis in Africa. Some African literary artists and dramatists especially of Nigeria extract have been able to champion the call for the depoliticization of oil on the African continent. Furthermore, Theater for Development projects have also been used to create environmental awareness among transnational oil companies, the government, and indigenes of oil-producing communities with the need to innovating ways in which the Niger Delta environment can be protected from continuous degradation. Two critical approaches employed by literary artists on politics and oil in Africa are Marxism and ecocriticism. A close reading of their text suggests that they are either interrogating materialist culture or examining the environmental situation of oil-producing countries in Africa. Some of their works are a fusion of these two critical approaches. Oil has dislocated the Niger Delta's middleclass. This has created class apartheid of the super-rich and the poor with almost nothing in the middle class. The dislocation of the middle class contributed to the rise in the crisis in African oil-producing countries especially the Niger Delta of Nigeria.

BIBLIOGRAPHY

Adam, S. (2005). World energy futures. In: J. Kalicki and D. Goldwyn (eds.), *Energy and Security: Toward a New Foreign Policy Strategy* (pp. 21–30). Washington, DC: Woodrow Wilson Centre Press.

Akinboka, C. (2011). Human rights to healthy environment. *Public and International Law Department, Faculty of Law* (2): 158–168.

Babatunde, A. (2010). Environmental conflict and the politics of oil in the oil-bearing areas of Nigeria's Niger Delta. *Peace and Conflict Review* Vol. 5 (1): 1–13.

Basedau, M., and Wegenast, T. (2009). Oil and diamonds as causes of civil war in sub-saharan Africa: Under what conditions? *Colombia Internacional* (70): 35–59.

Billion, P. (2013). The political economy of resource wars In: J. Cilliers and C. Dietrich (eds), *Angola's War Economy: The Role of Oil and Diamonds* (pp. 22–29). Pretoria: Institute for Security Studies.

Binebai, B. (2015). Writing from below, voicing voiceless voices: A theoretical scrutiny of Niger Delta drama. In: S. Ejeke (ed.), *Theatre Business* (pp. 287–304). Abraka: Delta State University Press.

Bromwen, M. (1999). *The Price of Oil: Corporate Responsibility and Human Rights Violations in Nigeria's Oil Producing Communities.* New York, London: Human Rights Watch.

Cornelli, A. (2012). A critical analysis of Nyerere's Ujamaa: An investigation of its foundations and values. *A thesis submitted to The University of Birmingham for the degree of Doctor of Philosophy.*

Gonzalez, A. (2010). Petroleum and its impact on three wars in Africa: Angola, Nigeria and Sudan. *Journal of Peace, Conflict and Development* (16): 58–86.

19 Julius-Adeoye, R. (2013). The drama of Ahmed Yerima: Studies in Nigerian theatre. *Leiden University Repository.* Retrieved 22nd January, 2019 from

https://openaccess.leidenuniv.nl/bitstream/handle/1887/20858/Introduction.pdf?sequence=5.

McGuffie, A. (2013). African educational film and video: Industry, ideology, and the regulation of Sub-Saharan sexuality. *A thesis submitted in partial fulfillment of the requirements for the Doctor of Philosophy degree in Film Studies in the Graduate College of The University of Iowa.*

Oil for nothing, multinational corporations, environmental destruction, death and impunity in the Niger Delta. (2005). Retrieved 10th March, 2019 from http://www.africaresource.com/index.php?option=com_content&view=article&id=46:chevron-oil-pollution-and-human-rights&catid=36:essays-a-discussions&Itemid=346.

Sudan: Oil and War. (1999). *APIC, Global Policy Forum.* Retrieved 10th March, 2019 from http://www.globalpolicy.org/component/content/article/198/32878.html.

Switzer, J. (2002). Oil and violence in Sudan. African Centre for Technology Study – Ecological Sources of Conflict Project Canadian Centre for Foreign Policy Development 1–19.

The Editor. (2001). Wanted: A Sudan policy. *The Washington Post.*

Vallins, G. (1971). Drama and theatre in education. In: J. Brown (ed.), *Drama and the Theatre with Radio, Film, and Television: An Outline for Students* (pp. 161–184). London: Routledge & Kegan Paul.

Victor, D. (2004). *Climate Change: Debating America's Policy Options.* New York: Council on Foreign Relations.

Yergin, D. (2006). Ensuring energy security. *Foreign Affairs* Vol. 85(2): 1–7.

Chapter 5

Through the Postcolonial Lens

Reading the Environment in Narratives from India's North East

Kalpana Bora Barman

"And in that gaping wound of the earth he buried the boar's tooth, the dismantled gun and Imchanok the hunter" (Ao 2009, 40). Noted writer Temsula Ao's volume of short tales titled *Laburnum for My Head* (2009) contains, among others, a story that is intriguingly called "Death of a Hunter." It narrates the story of Imchanok, a famed hunter known for his prowess and skill with the gun. Commissioned by the authorities to kill a vicious wild boar that has been wreaking havoc in the village, Imanchok readies himself for the task even as he remembers an earlier encounter when he shot a wild elephant only to be robbed of his sense of accomplishment post the kill as he realized that *he* was placed "at the center of the eternal contest between man and animal for dominion over land" (29). In killing the boar Imcha's confidence as a hunter, derived from a deep-rooted construction of the hyper male and the associated ideas of bravery and courage, is challenged by the dead animal whose act of dying reminds him of his insignificance in the larger scheme of nature and her maneuverings. And therefore, in an attempt at redemption, he buries a tuft of his hair along with a tooth of the boar and his now-dismantled gun, thus underscoring the title of the story and more importantly, the premise of man's interpersonal and intrapersonal relationship with nature and her kind.

Imcha's story may perhaps be located as the entry-point into the larger debate of the man-animal (and by extension the environment) relationship that is fraught with power-play, destruction, and dominance. Contemporary discourse on the environment has gained currency through the rise of ecocriticism as a discipline of study within the humanities largely due to the growing awareness of the irreparable destruction man's over-arching ambition has wrecked on our planet. Writers and activists are now being increasingly vocal about the need to let nature heal. Amitav Ghosh, for instance, has continued to engage with the ramifications of the man-environment relationship in his many works of fiction and nonfiction. In *The Great Derangement: Climate Change and the Unthinkable*, Ghosh reminds us of the "uncanny intimacy of our relationship with the non-human" (2016, 43) and also laments the collective failure of writers to represent a holistic and organic picture of this relationship primarily because representational acts have restricted to the individual and avoided the voice of the nonhuman. One may be reminded of Rudyard Kipling's *Kim* (1901) that famously described India as "grey" and "formless," thus depriving the land of any structure or Joseph Conrad's *Heart of Darkness* (1902) where the terrain is described as "featureless," "empty," and a "wilderness." Both narratives tend to oversimplify the representation of two very distinctive geographical locations by choosing to gloss over the distinctive geopolitical features of the land represented. Such representation is not only restrictive but also potentially thwarting to any true representation of these terrains, given the hegemonic authority of canonical texts in academia and the politics of writing itself that is enmeshed in the complex matrices of power, ideology, and affiliation. Anthony Carrigan, in his essay titled "Nature, Ecocriticism, and the Postcolonial Novel," echoes Edward Said's famous assertion that imperialism was premised on "an act of geographical violence" (2016, 81). Carrigan further argues that nature itself is a contested term whose universality is compromised by the many local narratives of the indigenous people that do not figure in the very Eurocentric construction of the term. Carrigan validates the plurality of nature by quoting Phil Macnaghten and John Urry that "there is no singular 'nature' . . . only a diversity of contested natures . . . constituted through a variety of socio-cultural processes" (2016, 82). In his Introduction to *The Cambridge Introduction to Literature and the Environment*, Timothy Clark asserts that "nature" is a construct of western thought and elaborates on its complexity:

> For an environmental critic, every account of a natural, semi-natural or urban landscape must represent an implicit re-engagement with what "nature" means or could mean, with the complex power and inheritance of this term and with its various implicit projections what of human identity is in relation to the non-human, with ideas of the wild, of nature as refuge or nature as resource, nature as the space of the outcast, of sin and perversity, nature as a space of metamorphosis or redemption. (2011, 6)

Clark argues that the role of sociocultural practices in the understanding of nature is hugely significant for this makes possible alternate readings of the environment that is conversant with the locale and the local thus rendering it an exercise in intertextuality. All texts, according to Dominic Head, carry a sense of "environmentality" and, therefore, it becomes imperative to examine the interactive exchanges between humans and nature specific to their demography, social life, and cultural practices, as Ghosh, Margaret Atwood, Jhumpa Lahiri, and Indra Sinha—to name a few—effectively do, to remind writers and readers of the urgency for developing a framework that is inclusive of the local realities of the land presented.

In the North East of India, environmental concerns have been the premise for much of the literature produced. The land of Seven Sisters (Assam, Arunachal Pradesh, Manipur, Meghalaya, Mizoram, Nagaland, and Tripura) and a lone brother (Sikkim), the "North East" is a seemingly all-encompassing term that is reflective of colonial idea of the north-eastern frontier prevalent in the 1970s. Contemporary rhetoric of the region is an off-shoot of this collective term and interestingly this rhetoric often designates the North East as a space that is conflict-ridden, tense, and characterized by insurgency and militant politics fueled by identity crisis and linguistic assertion. Its shared international borders only strengthen its status as a reserve of multicultural practices pertaining to everyday life and living. The North East also boasts of space of commonality in terms of its shared cultural geography, rites, and rituals. This geographical region has always been seen as separate and starkly different from mainland India, but it is replete with tales of political instability, ethnic clashes, resilience, and self-assertion. The ecology of the region, with its lush greenery and unruly hills, is a compelling force that gathers in its ambit the many differences—political, social, linguistic, and cultural—of its people. The many myths and legends testify to the shared ecological legacies by the inhabitants and showcase an alternate narrative of the otherwise prevalent human-environment dichotomy in a distinctive manner. The framework of postcolonial ecocriticism defined as the "relationship between literature and the physical environment" (Glotfelty and Fromm 1996, xviii). It is pivoted on the relationship between man and nature but recognizes the environment as a mere passive setting for such interaction. In his essay "Environmentalism and Postcolonialism" (2005), Rob Nixon proposes the possibility of bringing environmentalism into dialogue with postcolonialism to explore the possibilities of an inclusive democratic ecology intrinsic to this paradigm. As such, the overwhelming presence of alternate narratives of the environment only legitimizes the inter- and intra-textual nature of the man-nature relationship. It makes impertinent the examination of these under the light of postcolonial environmentalism that challenges and dismantles existing tropes of power and exploitation of the environment to make way for narratives of coexistence premised on mutual respect and an acknowledgment of the significance of the nonhuman and a shift of perspective from an anthropocentric to a biocentric that would "affirm the intrinsic value of all natural life and displace the current preference of even the most trivial human demands over the needs of other species or integrity of place" (Clark 2011, 2).

In the context of the present chapter, the premise of postcolonial environmentalism may be traced in the writings of Mamang Dai's *The Legends of Pensam* (2006) that reveal a deep understanding of the man-environment dynamics of the Adi tribe in Arunachal Pradesh. The text seamlessly merges myth, history, folklore, and fiction central to the tribe to highlight the ways in which the Adis locate nature at the center of their existence. The works of Temsula Ao and Dhruba Hazarika, among others, too draw heavily on the nuances of human-environment relationship. Temsula Ao has continued to engage with and examine the sociocultural politics of the region in widely anthologized works such as *These Hills Called Home* (2006). *Laburnum for My Head* published in 2009 is a celebrated work where Ao seeks to intervene into the traditional ecological paradigm to pave way for new understandings of contemporary existence, one that is in tandem with the ecosystem that makes possible the coexistence of man, animal, and the environment. The title story of the volume, "Laburnum for my Head" narrates the story of Lentina, an extraordinarily strong-willed widow whose greatest desire was to have some luxurious yellow laburnum trees in her garden, "She had always admired these yellow flowers for what she thought was their femininity; they were not brazen like the gulmohars . . . . The way the laburnum flowers hung their heads earthward appealed to her because she attributed humility to the gesture" (2009, 2).

Repeated attempts to grow these trees resulted in failure until, on the day of her husband's funeral, she decided to have a laburnum tree planted on her own grave at the time of her death instead of the customary headstone. Knowing that her decision would be met with ridicule and disapproval from her immediate family, she worked in secret with her driver, Mapu who, compelled by "dutiful service," conspired with Lentina to turn this wish into reality. From selecting a plot in the land to negotiating with the authorities to purchase the same, Lentina resolutely went about her task, not heeding her family who believed that their mother had lost her mind. Lentina purposefully decided to leave out her sons from her mission. Once the laburnum saplings were planted the final paragraphs of the story narrate the old lady's excitement as she waited for the blooms, her ecstasy when she witnessed her plants grow, and the strange workings of nature for Lentina was now old and ill, but the sight of laburnum plants seemed to infuse her fatigued body with vigor. With her death ends the "story of the un-dramatic life of an ordinary woman" who wished to see the "glory of buttery-yellow splendor" bloom at her grave. This "extraordinary" event that occurs every year in May with the laburnum tree "bedecked in its seasonal glory, standing tall over all other plants, flourishing in perfect co-existence, in an environment liberated from all human pretensions to immortality" (2009, 20). This reminds the reader of a life that exists in tandem with nature, a nature that "does not possess any script, abides there: she only owns the seasons" (2009, 2). Dhruba Hazarika's novel, *A Bowstring Winter*, resonates with this sentiment when Charley proclaims, "I'm of the hills, and the hills are in me" (2006, 118). Ao's narrative reverberates with a similar sense of reverence for nature; in *These Hills Called Home*, she writes of "youngsters of today who have forgotten how to listen to the voice of the earth and the wind" (2006, 32). Lentina's wish, thus, is reflective of the larger human need to remain connected to one's roots through the happy cohabitation of man and nature.

Lentina's sense of purpose is mirrored in Imchanok's story in "Death of a Hunter." Imanchok recounts his earlier encounter with a rogue elephant that had destroyed fields, homes, and lives. He had been commissioned by the Deputy Commissioner for the task, for everybody was of the opinion that none but Imanchok possessed the prowess and cunning to kill the elephant. Equipped with elephant-shooting rifle and ammunition, the hunter readies himself even as he considers the difficulty of the task,

> What do these sahibs know about the jungle? Do they think the elephant will be waiting at a convenient place for me to go and shoot him? Don't they know how intelligent these animals are, that they can almost think like human beings? And the area they can cover when they decide to run? (2009, 24)

Despite such thoughts, he complies with the government's orders—refusal may lead to the suspension or revoking of hunting licenses. Yet, what is important to Imcha is not the threat to his profession but the threat to his reputation—failure to kill the elephant may cost him his hard-earned name. He begins preparing for the hunt in earnest, enlists the help of his most trusted assistants, and engages in discussions and strategies. As they take their position in the jungle after digging a huge hole for the elephant to fall in, Ao delves deep into the mind of the hunter as Imcha battles his fears and waits with his team in the depths of the jungle. When the elephant appears, the storyline changes focus and present the elephant in all its individual grace. Unaware of death lurking nearby, the elephant appears "calm and serene" even as it enjoys a dust bath. Imcha is fascinated by the animal, for now the "initial terror of the unknown was relieved by the spectacle they witnessed" (2009, 27). The day begins to grow hot and Imcha wastes no time in shooting the elephant. The animal is stunned by the attack and Imcha watches in "awe and fascination" the as the huge animal is caught in the slow dance of death. Divested of his power and strength, the animal watches Imcha helplessly, its eyes conveying a message, or perhaps asking a question. Unlike previous times when Imcha would hunt according to his wish, this time his prey was "allotted" to him. Ao locates this instant as Imcha's moment of reckoning: his glory as a hunter has been compromised, for *he* was placed by an authority other than him. The elephant hunt won accolades, name, and fame for Imcha but he was a changed man now—haunted by the images of the dying elephant, he resolved never to hunt again. He remembers an earlier time when he had killed a monkey that had displayed great leadership as it tried to herd out its mates as they were cornered by Imcha and his team. Filled with remorse, he instructed his wife Tangchetla not to use grain from the hut where the monkey had been killed. He did not want to "contaminate his main barn by bringing in paddy soiled by a pack of monkeys and tainted with the blood of the leader" (2009, 33). In the present moment when Imcha shoots the boar, the bullet finds its mark but Imcha walks away, unable to bring himself to witness the animal's death. He soon takes ill. Haunted by the spirit of his kill, a guilt-ridden Imcha, goaded by his wife, returns to the jungle and performs a simple ritual so as to honor the life he had ended, and thus exorcise himself of his deed. He locates the spot where the carcass of the boar had died, picks a tooth of the animal, and washes it clean. He then goes to the spot where he shot the boar, tears a tuft of his own hair, blows it toward the jungle, and leaves without looking back. Once home, he buries the boar's tooth along with his now-dismantled gun, thus bringing an end to his journey as a hunter. This simple act of seeking nature's forgiveness forms the crux of the story. "Death of a Hunter" is not about the Imcha's death; it is about the death of the hunter in Imcha who realized the futility of his act, the meaninglessness of his power and skill that kills rather than protects. More importantly, the story underscores man's oneness with nature, his interpersonal relationship with it premised on interdependence and the respect and dignity that all human beings and animals deserve.

"Flight," the final and shortest story of the volume, once again underscores the man-animal relationship through the tale of a boy and a butterfly. In a clever role reversal, the story is narrated by an insect, the caterpillar, rather than a human being, and this lends the tale a very interesting perspective:

> Life began for me in the wide-open spaces of a vast cabbage field, in fact, on the underside of a big leaf near the passage-ways between the rows of plants, which crisscrossed the entire length and breadth of the field. From the minuscule speck of seed left by the flitting mother, I slowly evolved into an elongated green form, blending in perfectly with the big leaf. (2009, 104)

However, the caterpillar is soon captured by Johnny, a terminally ill boy, and kept in a box; the "wide-open spaces" of the opening paragraph are immediately replaced by darkness as the insect lay captive:

At that instant, my former life of wide-open spaces and bright sunshine vanished, and the new one of intermittent light and darkness began. Light, when he opened the lid to peep at me, and darkness again when he lowered it. The periods between light and darkness were regular to begin with, but as time went on, they became longer and longer. Sometimes days would go by without a glimpse of light. (2009, 105)

As the caterpillar metamorphoses into a beautiful butterfly, its young captor slowly inches toward his death. Ao employs the metaphor of flight to trace the parallel journeys of the captor and the captive—one reaching out to life and the other spiraling don the vortex of death. Johnny's refusal to recognize the beauty of the newly formed butterfly—"Beautiful? Dragon, what happened to you? You look ugly."—does not deter it from trying out its new wings and leave its dark prison. The butterfly knows that it must take flight, just as it knows that it has to leave behind Johnny's "dying universe" (2009, 106). And so "I flapped by wings and was soon fleeting away without a backward glance" (2009, 107).

Temsula Ao's engagement with the nonhuman element in the environment mirrors the complexities of environmental exchanges, an exercise that finds expression in Dhruba Hazarika's *Luck*, which is another nuanced portrayal of this relationship. Set amid the backdrop of everyday life and living in the North East, the author portrays the trials and tribulations of man but chooses to center the tales around the birds and animals that populate the tales. In fact, most stories in the volume locate the bird or animal as the pivot around which the narrative is centered. In "The Hunt" a young doctor, unable to come to terms with the bestiality of his cruel act of shooting the pregnant doe, refuses to cut her open—"I can't do it. I can't do it" (2006, 4)—because inside the dead doe lay "tiny spindly-green, almost transparent bodies" (2006, 3) that remind him of his own wife who died at childbirth a year ago. When Adrian and Nataram return after burying the unborn fetuses they find the doctor "lying on top of the slain doe, holding her neck. His clothes were damp with blood and in his hand he held the bullet that he had fired a while ago" (2006, 4). The trio then leaves the forest with Adrian looking "into the night" waiting "for the forest gods to forgive us" (2006, 4). This guilt of having trespassed the sanctity of the forest and wrecking the order of nature is beautifully elaborated in "The Leopard" even as Bahadur, Dilip, and the narrator climb the Blue Mountain to look for Bahadur's cow that had gone missing. They soon discover the dead cow, killed by a predatory leopard that had been prowling in the area. Even as they recounted their tale to their friends, the trio could not overcome the strangeness and "terrible splendor" of the leopard that had jumped out of the dead cow's belly when they had neared it. The story ends on a note of pain and what may be described as the revenge of nature for soon after the leopard had been killed by the villagers two starving leopard cubs were discovered beyond the Blue Mountain. Any sense of terror that the reader feels on reading about the leopard's hunt of the cow is almost immediately replaced by a sense of redemption and justice served, although the justice met is tinged with a hint of sadness at the fate of the two leopard cubs. Hazarika refuses to give the reader time to grapple with emotions; the stories are short and crisp so that even before the eventuality of the narrative sinks in the author immediately brings in another emotion, an element of surprise to completely unsettle the reader.

In yet another compelling narrative titled "Chicken Fever," the protagonist Rattan Deb Barman's sense of inadequacy, made more pronounced by his beautiful wife, is complimented by the hen-coop and its inhabitants: he built the coop himself, and "gave it time and attention, at fixed times of the day, much as army men polish their shoes and iron their uniforms at sunrise" (2006, 34). The hens and a rooster bring a sense of order in Rattan's life so much so that one day when a mongoose disappears with the rooster's neck in its jaws, Rattan is infuriated and vows to kill the animal "so that you will know the pain you have given me" (2006, 37). He soon leads a team of men on a police raid, although reluctantly, to destroy the illegal settlements of illegal immigrants that had mushroomed in Midlicherra but is plagued by self-doubt:

And why are you so sacred, Rattan Deb Barman, when Basumatary sits calmly smiling next to you? You were not born for courage, Rattan Deb Barman. You were born to cringe and live off others' fat and toil and bravery. So why do you pretend? You can still halt the men, say you have cramps and turn back. (2006, 43)

Ashamed of this thoughts Rattan goes about his work, instructing his men to destroy the huts and haystacks that were bereft of people who had supposedly fled earlier. Rattan thought they "could melt into the jungle as if they never existed" (2006, 45). The sight of a black fat hen and a couple of chicks soothed his frayed nerves and he could "smell the musty-earth, feather-rich chicken smell that he loved so much" (2006, 49). Even as he whispered to the hen, "Don't be scared of me, fat mother hen" (2006, 49) he was engulfed by a sense of unaccountable happiness, the hostility of the place melting away with the sight and smell of the familiar. The hen symbolized "hope" and Rattan felt less desolate and afraid. Despite the imminent danger, the hen stayed on with her chicks, protecting them and even the "sound of things being smashed and pummeled around her hadn't driven her away" (2006, 50). Inspired, Rattan does not disclose his discovery to Basumatary—"Anything he said would have sounded awkward and probably embarrassed him" (2006,

50). He picks stacks of hay to cover the hen but immediately discovers a young girl hiding amid the haystacks; she must have probably been left behind by her people in their hurry to escape. The barely clothed girl waited for Rattan to make his move—the author here clearly hints at the atrocities meted out to women at the hands of men. Accounts of army-militant clashes in the North East are replete with episodes of molestation and rape of helpless women who become victims of the struggle between the conflicting forces. Rattan too is momentarily lured by the raw beauty of the frightened girl but that momentary weakness soon gives way to the greater emotion of wishing to protect her from the men accompanying him, for Rattan too was aware of the plight of such women caught in the midst of conflict: "He put a finger to his lips to silence her and picked up the second bundle of straw and pressed it around" hoping that "she should be able to breathe though the straw until they were gone" (2006, 52). Meanwhile, the hen and her chicks are soon discovered and while the captor holds up the chicken in a "lust for good food" (2006, 53), Rattan is filled with blind rage and roars at the man to release it immediately. Rattan battles his own inherent lust as he is caught between the hunger for the chicken and hunger for the girl hiding amid the hay. Praying that "the girl wouldn't move, would hold her breath and still her wildly beating heart" Rattan orders his men not to take anything from this ravaged land (2006, 54). In order to ensure that the settlers do not come back, the men set fire to the huts even as Rattan walks back to his jeep "past the blackened field and the smashed bricks, past the broken and gutted huts, past the plantain and papaya trees, past the breathing of invisible people behind the trees, past the whispers of missing women and children" (2006, 59). Hazarika chooses this moment to converge the narratives of Rattan, the hen and the girl to draw attention to the ways in which each element is intrinsic to the organic function of nature. He knew "it was the black hen that brought us good luck. It was good luck. Very good luck" (2006, 60). And this perhaps is the reason why he saved the girl, "You'll live, and one day you'll be a mother, like that fat, brave hen. You'll live, and make life. And perhaps you'll remember me" (2006, 60). This is Rattan's moment of self-discovery, a moment of unlearning that perhaps is necessary to relearn the ways and workings of the universe premised on inclusivity and recognition of the significance of the not-so-vocal participants.

"Ghostie"—the tale of the ghostlike neighborhood dog—reflects this idea even as Jycbo, Deep, and the narrator become "cruel as only children can be" (2006, 69). Subject to multiple tortures by the three boys, Ghostie steadfastly holds his own against his perpetrators by refusing to display any hint of a reaction and chooses to remain a silent victim even as it is stoned, hit, or splashed with boiling water. The eerie silence of the deeply wounded dog unsettles the boys, its "aloof and confident look, the shine of his coat, the broken ear, the undulating muscular body delineating what each of us hoped to be one day: tough both inside and outside. Though even as boys we knew he was everything that we would never be" (2006, 76). They decide to ignore it but not for long. Stricken with remorse after his unspeakable deed, the narrator—for he was the one who splashed Ghostie with boiling water—ensured that the dog was given the leftover food from his mother's kitchen. He admired Ghostie's resilience and believed that Ghostie was not a ghost but "the spirit of final justice" (2006, 77). Ghostie's refusal to display any reaction or pain baffles Sanjay, the new boy in the neighborhood. To prove his friends wrong Sanjay shoots the dog repeatedly with an air-rifle, "I saw Ghostie's head jerk back as the pellet entered his eye—I think the right eye. The body quivered in silent protest. He came groggily to his feet" (2006, 79). Even as the narrator punches Sanjay for his cruel act, Ghostie bleeds and slowly settles down on the ground, his head on his paws. The narrator is baffled—Ghostie epitomized silent strength, a still courage and dignity that would not be shaken easily. Why then "had Ghostie let himself be hurt? Why had he shown me that he could be hurt?" (2006, 80). Now that it cannot search for food on its own, the narrator ensures once again that Ghostie is never hungry and continues to bring him pieces of meat and bread until the wounded dog slept "a blameless sleep" (2006, 81). Ghostie's resilience underscores man's powerlessness over animals but also shows how animals may become instrumental in bringing about a change of heart. The animal world finds a voice against the ill-doings of man through Ghostie and reminds the reader of their ability to choose their responses that may vary drastically from human expectations. "Asylum" elaborates on this point when animals respond to human extremities in their own unique way. Hargovind, the vet-cum-psychiatrist, shares a close bond with animals so much so that every day he carries over leftover rice and dal from the asylum to feed his pigs, cat, and dog. In order to cover his illicit relationship with Reema, Dhaniram Dutta's wife, he regularly injects the latter with a hallucinatory drug to ensure that he remains in the asylum. However, when Pusso, the cat, and Missus Tippoo, the dog, suddenly become violent, Hargovind is baffled. The otherwise silent pigs too suddenly go out of control and begin to fling themselves against the walls of their sty. Meanwhile, Dhaniram suddenly begins to exhibit normal behavior and walks away from the asylum, as do his other three patients. When Dhaniram reveals to Hargovind his knowledge of the affair and his ways of arriving at a cure for his condition, the latter vents his anger and frustration on Missus Tippoo who attacked him fatally:

> Missus Tippoo fell back on the floor. In a quick move, Hargovind brought the barrel of the gun down on her head, again and again, until she stopped moving

altogether. There was blood on the kitchen floor and Missus Tippoo's once beautiful, graceful body lay mangled in a mess of fur and blood. (2006, 108)

This act of violence only signals Hargovind's spiraling down into insanity resulting from guilt combined with fear for his life. As the animals become erratic and the pigeons die one by one, the story moves toward a closure with Manjulla, Hargovind's assistant, lamenting Hargovind's misplaced trust in animals.

"Vultures" carries forward the sense of foreboding as once more man's fate is inextricably linked with the animal world. "Wide of wingspan, dark and noiseless, [the vultures] were like etchings against the sharp blue sky," says the narrator (2006, 113). Fascinated by these creatures, the narrator together with Lien-Thang enters the woods and spots eleven of them, "And they were quiet. For a second or two I wanted to run. Huge and ominous, they looked like ogres. Black and grey and hunched, their necks old and crooked, their legs scrawny but sturdy, they formed a loose circle, unmindful of our approach" (2006, 120). The narrator, however, fails to fathom these birds. For him, birds were the gentler species, "pigeons and sparrows and mynahs and parrots that lived on gram and green grass and the occasional earthworm. Birds were crows, despite their never-ending cawing, and birds were owls, eternal in their dignity and wisdom. I could not think of vultures as birds" (2006, 115).

Yet, he was captivated by these "disciplined workers" as they go back into the forest after a day of "honest labor" (2006, 122). Hazarika juxtaposes gentleness and brutality perhaps to underline the possibility of happy coexistence of opposing elements. Humans too possess similar conflicting qualities but perhaps they lack a sense of regimented order as they go about their daily lives. In "Soul Egret," the narrator considers the egrets to be his "possessions" and "often leaned forward from the edge of my verandah and strained my neck tom catch a glimpse of the fluffy feathers and tiny yellow legs" (2006, 124). He scolds the children from the quarters who often climb the fig trees and disturb the birds, but they only grin and mumble something about "the taste of bird's flesh" (2006, 125), immediately transporting the reader to the embedded cruelty in Ghostie's story. The children's attitude makes the narrator fiercely protective of the birds; he sis soon heartbroken when, on a particularly stormy night, as many as sixty egrets lost their lives and the children excitedly collected the dead bodies the next morning in the prospect of a sumptuous meal. The relationship between the narrator and the birds is sealed when many days alter the narrator catches hold of an egret as it tried hiding under the DM's car. He brought it "close to my chest. The wings beat at my neck and shoulders briefly, then the bird was till against my heart" (2006, 127). As the bird attempted to fly away, "I dipped my head and kissed its neck" (2006, 128). Later, "the brief flutter of wings against the bare skin on my neck" continued to remain a singular memory for the clerk, signifying the essence of the unspoken bond he shared with the bird.

The title story of the volume is called "Luck" and centers on a lonely man and a solitary pigeon. The narrator has been trying to keep pets for many years, but every time his efforts result in failure as the animals and birds either run off or fly away "like guests who had been forced into being guests, or people who had strayed into camps that cut off their freedom, they stayed awhile, enlivening the compound, and then, when the spell came, they were gone and there was nothing you could do about it" (2006, 6). The narrator almost began believing that their house was taboo for animals until one day a pigeon appeared out of nowhere and, even after many days, stayed on. He was named Luck. Unlike the previous ones, Luck had "sparkle and confidence.. In Luck's "sparkle and confidence," "his proud gait, the chest always thrust out, as he came to pick the gram from my hand; the way his eyes darted around without fear or suspicion; the graceful fall of his wings as he settled like an owl over the box for the night. Luck was the man of the house; there wasn't any doubt about it" (2006, 14–15). Luck soon becomes the narrator's constant companion, engulfing him with a blanket of trust and security. The sense of togetherness that these two completely different beings derive in each other's company is, once again, a pointer to the complexities of the man-animal/bird relationship. Luck's story is a happy reminder of an essential truth: man's relationship with nature is not premised on cruelty and exploitation alone. Love, trust, and companionship too form the basis of these association and, more often than not, these qualities overwhelm the otherwise purported narrative of the man-animal world.

DeLoughrey and Handley, in their Introduction to *Postcolonial Ecologies: Literatures of the Environment* locate the work of Edward Said to elaborate on postcolonial ecology and highlight the significance of place in the representational exercise:

> A spatial imagination made possible by the experience of place. Place has infinite meanings and morphologies: it might be defined geographically, in terms of the expansion of empire; environmentally, in terms of wilderness or urban settings; genealogically, in linking communal ancestry to land; as well as phenomenologically, connecting body to place. (2011, 4)

It is also important to enter into a holistic dialogue with nature and the environment to make possible the emergence of untold narrative that may contribute significantly to the dialogue of postcolonial environmentalism. The tales discussed serve as examples of the ways in the untold and unrepresented may be given voice and centrality. While celebrating the multiplicity of the man-animal world, the tales make a clear and conscious digression from the prevalent modes of postcolonial representation of nature to underscore the centrality of nature as it is, and not as it should be understood. Such literature is thus infused with a sense of the local that belies the greater discourse of nature conversant with the history of the empire.

The literature of the North East is infused with an awareness of the cultural and ecological harmony. To a larger degree, writers of the region are driven by a conscious need to effectively represent the pulsating cultural ethos of communities and the connections with the environment. The stories in this chapter engage predominantly with ideas of the other, and locate the nonhuman at the center of the narrative, thus subverting the practice of marginalization of the minority. The narratives examine the spatial location of individuals in diverse relationship thus being thematically united. They record individual responses and reactions to everyday experiences and the nonhuman to reaffirm the selfhood of the subject premised on the presence of the other. New readings of the environment must therefore consciously do away with ideological hegemony and engage with the elements of the earth in all its totality, taking into cognizance the remarkable diversity, embracing the differences so that a new narrative may be created that advocates peaceful coexistence for all.

BIBLIOGRAPHY

Ao, Temsula. 2006. *These Hills Called Home*. New Delhi: Penguin.
———. 2009. *Laburnum for My Head*. New Delhi: Penguin.
Carrigan, Anthony. 2016. "Nature, Ecocriticism, and the Postcolonial Novel." In *The Cambridge Companion to the Postcolonial Novel*, edited by AtoQuayson. Cambridge: Cambridge University Press.
Clark, Timothy. 2011. *The Cambridge Introduction to Literature and the Environment*. New York: Cambridge University Press.
DeLoughrey, Elizabeth, and George B. Handley. 2011. *Postcolonial Ecologies: Literatures of the Environment*. Oxford: Oxford University Press.
Ghosh, Amitav. 2016. *The Great Derangement: Climate Change and the Unthinkable*. New Delhi: Penguin. Glotfelty, Cheryll, and Harold Fromm, editors. 1996. *The Ecocriticism Reader: Landmarks in Literary Ecology*. Georgia: University of Georgia Press.
Hazarika, Dhruba. 2006a. *A Bowstring Winter*. New Delhi: Penguin.
———. 2006b. *Luck*. New Delhi: Penguin.
Nixon, Rob. 2005. "Environmentalism and Postcolonialism." In *Postcolonial Studies and Beyond*, edited by Ania Loomba, et al. Durham, NC: Duke University Press.

Chapter 6

"Aesthetics of Belonging"

Construction of a Postcolonial Landscape in Daud Kamal's Poetry

Humaira Riaz

Environment, aesthetics, belonging, and postcolonialism is an arrangement integrated in one united whole in the contemporary world. Environment is the natural world that offers human being portrayals to be appreciated or denounced subject to their sense of appraisal usually defined as aesthetics. Aesthetics refers to artistic exercise of exploring oneself; a self-realization and consciousness of the surroundings. Aesthetics is the examination of subjective and emotional values. Aesthetic professes provision of sensory pleasure rather than sentimentality. All objects perceived as beautiful are important in the sense that these represent our values and values in turn are beautiful as these pave our actions.

If aesthetics is defined as sensibility to appreciate the work of art, then aesthetics of belonging is appreciating and critically being conscious of the environment and natural objects around us. Aesthetics is to understand "nature" in its existing condition unlike a historical or popular understanding where "moon" represents "beauty" and "flow of river" demonstrates "life." It means a thing of *beauty* may offer joy forever, provided we seriously consider that description of beauty remains with the spectator. To my understanding, aesthetics basically exposes sensible experiences relevant to arts and scenic nature and their appraisal. The more its horizon expands, insightful the critical appreciation becomes. Conventional approach to aesthetics defines it as a technique to identify and illuminate perceptual experiences. However, contemporary approaches may not consider it instrumental in celebrating sensory experiences only. Rather, it involves a sensible celebration focused on positive as well as negative values.

A work of art and literature is likely to produce an aesthetic effect if the work implies intensity of experience and contains emotional involvement of the writer. Such work reflects newer artistic expressions depicting specific culture and conditions. Such conditions pertain to the concept of belonging, a major concept in postcolonial literature. It arises from alienation, ambivalence, and displacement as a result of territorial occupation, in the stricter sense characterized as colonization. Colonialism has a complex history of exile and dislocation. Independence Movement in subcontinent achieved its aim with the independence of Pakistan and India on August 14 and 15, 1947, respectively. Independence of India was considered by Conservative English a moment when British empire lost its power over the world. British colonial empire changed the cultural as well as natural landscape of Indo-Pak subcontinent. Colonial literature for decades had projected the landscape implying colonized inferior in relation to the superior colonizer. Further, postcolonialism embarked as a dominant reactionary force to record protest and resentment to colonial attitude and approach.

Postcolonial writers did not consider men as "blank tablets on which the environment inscribes a culture, which can readily be erased to make way for a new inscription" (Berry and Murray 2000, 13). Therefore, the sense of belonging to origin remained a consistent power in nourishing or repressing people's attitudes. A fruitful dialogue with the past enabled one to renew the "fossils," which were buried long before within the hearts and were equally a part of one's ancestors (Ashcroft et al. 2006, 187). Place, therefore, remains a constant reminder of separation from one's roots in postcolonial studies. It is the journey to other land "a journey into the wilderness" (Atwood as cited in Ashcroft et al. 2006, 395).

A similar journey was caused by the historical trauma of Indo-Pak subcontinent in 1947, which echoed individual and "collective memory" (Saint 2010, 10). The situation changed the entire landscape of "recollection" (Saint 2010, 10). Works of art and literature reproduced the sufferings of the survivors. Crude ideologies based on religion, caste, creed, and culture played a vital role during the decolonization of Indo-Pak subcontinent. For the reason, early writings during 1940–1950 reflected high sentiments and emotionalism as the writers were directly affected. Therefore, the response was tragic and poignant to the events they witnessed. However, there was a shift in response during 1960 till date. New perspectives were included by postcolonial writers who reflected on the events prescribing new dimensions to the affected in the aftermath of decolonization. Colonial perceptions of partition were founded majorly on communal and religious differences. However, postcolonial writers unlocked their emotions to show resistance to the colonizer and the process of decolonization with varying perspectives.

To carry forward the discussion, we see partition and its continued impact have revitalized an understanding of postcolonialism in the present-day Pakistani English literature. Contemporary Pakistani English poets seek to establish "aesthetics of belonging" by means of language and literature. This chapter presents the argument that nature, as a place to break free, a revival of the concept of heaven, is a colonial visualization used as defying colonial powers to represent post-partition Pakistani landscape. My reading of renowned Pakistani poet Daud Kamal's poetic works explores how he constructs postcolonial landscape by exercising visual representation challenging colonial models. I also attempt to expose how nature situates in various moments to identify identity crises in Pakistani society. The study also reflects upon the complexity of relationship between environmental destruction and postcolonialism through Pakistani English literature dominated by a high sense of belonging and identity. The relationship between earth and human being described as a means to respond to the disintegration of history by "conceiv[ing] totality but willingly renounc[ing] any claims to sum it up or to possess it" (Glissant 1997, 27). "Aesthetics of belonging" for Glissant expands to express appreciation of beauty even when colonial aggression has destroyed the land and sea (Glissant 1997, 27). Following Glissant, in my reading of Pakistani English poet Daud Kamal, I trace how poetry makes evident the power relations that constructed nature as anti-empire, disconcerts culture, and reinstates a consciousness of the cultural transformation that regulated their political culture in the postcolonial epoch. Concept of belongings is integrated into environment.

We all die
even those
who've not yet lived. (Kamal 1997, 26)

Daud Kamal, a distinguished postcolonial poet translated Urdu poems of renowned poets Mirza Ghalib and Faiz Ahmed Faiz into English. However, his original works are in English language. His poems published in different anthologies reflect great economy of words and precision to uncover a universe of meaning. Undoubtedly, his works are of an imagist who captured universe in a single illustration. *The Compass of Love* (1973), *Recognition* (1979), *A Remote Beginning* (1985), and *Before the Carnations Wither* (1995) contain his original poems in English.

Daud Kamal is famous for recording his protests and resistance to British colonizers. Greatly influenced by modernist poets Ezra Pond, Yeats, and Eliot, Kamal wrote mostly in English language. His English translations of Faiz Ahmad Faiz Urdu poetry hold a remarkable place in Pakistani English literature. Kamal's poetry reflects belonging, home, identity, and a sense of loss in the milieu of partition. His poems construct a postcolonial landscape through visual representation challenging colonial models. For the present study, I have selected poems from two volumes to explore how deeply his words reflect mind of a postcolonial writer whose brain and heart ache for the belongings in the backdrop of colonization. His poems "The Rebel," "Floods," and "Rain on Moss," are selected from the collection *A Selection of Verse* (1997) and "Anniversary," "Kingfisher," and "Winter Rain" are taken from anthology of his poems *Rivermist* (1992).

Overall, his poems present a scattered, fragmented multidimensional symbolism, which relates present to the past and future. His works are deeply engrossed in natural environment in the context of resistance within the homeland. Interestingly, the titles of his poems are in capital alphabets, which in the first place accentuate the environmental aspect of the poetic works. Furthermore, these explain Kamal's attachment to nature in its pure, contaminated, (or ruined in the colonial context) as well as obscure form.

RESISTANCE, IDENTITY, AND AESTHETICS OF BELONGINGS

Resistance has always been a discernable subject with postcolonial writers. The writers belonging to the colonies use their pen to bring forth the psychological disorders generated by colonialism. It is defined "as a theoretical resistance to the mystifying amnesia of the colonial aftermath revisiting, remembering, and interrogating the colonial past" (Gandhi 1998, 4). Postcolonial writers emphasize concerns of human identity and sense of belonging in a colonial and postcolonial world. They are the disappointed colonial individuals investigating their sense of belonging and identity, which dissociated from the process of colonization in an intriguing way. Daud Kamal is a conventional postcolonial writer whose works reflect colonial and postcolonial dilemmas of Pakistan.

Identity and belonging in Kamal's works has a strong relationship. Both are vital "human rights." Identity is "self-actualization" and representation of other individuals. One's identity is therefore, relational, that is, understood in relation to other individuals. Kamal constructs social and political identity of the individual in his poem "The Rebel." The rebel is understood as an insurgent individual resistance to "they." "Dawn," an important image representing the sun rise, is symbolic of life. However, in postcolonial world, the image turns contrary and signifies death; death chosen for the rebel by others. "Orchard" is the context for place where individuals are shown colors and fruit of life and afterward they are deprived of what nature has bestowed on them. Mingling nature with brutality of colonial power implies the helplessness of "nature" when instead of showering life rays; it is bringing death and desertion to people. "The Rebel" demonstrates agitation of the speaker in a land, which was separated for a cause. However, the existing situation appears to have reversed the roles of colonizer and the colonized. Deeply influenced by post-modernist imagery, Kamal's poem can be viewed as a work of an imagist or reductionist. The poem depicts an image of the "orchard" and associates the beautiful image of nature to death. Nature in this poem is anticipated as the "murdered" and "plundered" forsaken body. The poem asks for more sacrifice, which raises question to the sacrifices offered for a free homeland, free of class, gender, and religious discriminations. Moreover, nature, as a place to break free, a revival of the concept of heaven, is perceived by colonial visualization. "Orchard," a garden of fruit, symbolically represents a doorway to heaven. Ironically, individuals are killed for their "rebellion" against the state powers. Garden is also symbolic of a hybrid space of nature and culture. It illustrates utopian narrative in a dystopian world of power and belligerence. Words such as "Pandemonium of crows" and "empty horizon" mock the selfish attitude of the postcolonial society, where empty and perplexed minds perform the role of spectators (Kamal 1997, 31).

Devoid of feeling the change that revolution has brought, Kamal elaborates sense of loss and belonging associated to the land people left after partition. Here is the grave concern of acceptance on part of the people who did not migrate and refused to accept those who migrated to their part of land. The poem also identifies the continuous journey "between roots and routs" (Nayar 2008, 17). Kamal refers to the ironic rebellion of people against their own natives. Metaphor for earth as "blind" suggests the history-old oppression colonizer exercised and transferred to the individuals. Nature thus is portrayed as anti-empire. The poem reinstates a consciousness of cultural transformation, which regulated a political culture as a legacy of the colonial epoch.

Colonialism consequently left legacy of environmental problems. Process of decolonization was an "arbitrary drawn zone of administration" causing colonized countries internal and political destabilization (Wood 2015, 1). It also demonstrates that colonial powers left enduring scars on the development of their former colonies. Divide and rule policy favored imperial designs and to a larger extent paved way for dictators to establish a powerful grip on nation. It happened so by simply filling the "power vacuum" in colonial territories after freedom (Wood 2015, 1). This kind of situation breeds civil war, resentment, and corruption. Even politicians grow "prisoners of power . . . . ..seeking . . . absolute control" (Wood 2015, 35). Kamal's work highlights "the environmental scars" in the form of unconventional images of natural worlds (Wood 2015, 1). He draws an analogy between "wheat ear on the stubble" and the rebel who shows resistances to power hierarchies in the newly born state.

"Floods" is another poem contemplative of the colonial aggression at its worst. As the title suggests, floods are never human or earth friendly. It may be regarded as nature's revolt causing loss of human life, destruction of property, damage to the crops, loss of livestock, and worsening health conditions owing to waterborne diseases. The title of the poem can be understood as an analogy to colonialism, which destroyed human life and their land like gushes of water exercising their power of destruction. Water otherwise is taken as a sign of life and regeneration, however, its extremity and overpowering capacity is harmful. In the same way colonial strategies to develop, civilize, and educate the colonial subjects and areas resulted in psychological as well as physical deterioration of people of those areas. The first line of the poem reminds the readers of the treachery and seizure by colonial masters, which took over and carried away human desires of prosperity. Colonial powers plundered and injured their bodies and souls. Foreign intrusion and suppression sicken both. Kamal refuses to forgive the colonial aggression portraying it as "blind rivers," which carry away everything in its furious and conceited gush. Word "treachery" in the poem exposes colonial control as an enemy, betraying the natives, and causing them psychological and physical troubles. In the aftermath of colonial aggression, partition and displacement became the biggest and adamant hazard causing food, shelter, and health problems (Kamal 1997, 37).

Further, he uses image of a buffalo, which is stuck in muddy puddle. Parallel is the situation of postcolonial world where human beings are caught helplessly suffering from identity crises, issues of settlement, and at the same time nostalgic for belongings. The sense of heavy loss prevails upon the entire poem. The poet recalls the time of migration and the existing situation of the migrants living in forty years old hut with no evident change in their condition.

Reference to "forty years" reflects pre-partition era indicating strong feelings of something lost and left behind (Kamal 1997, 37). It depicts a bonding based on shared territory, history, language, culture, doubts, socioeconomic interests, and religious and ethnic links. Entire assets of the migrants consisted of "wooden boxes" and one sack of rice for survival at the time of migration (Kamal 1997, 37). Word "dowry" employs their position analogous to daughters in Asian countries. Dowry is given to help set up new household, especially where it is rare for a woman to work outside the home. It anticipates a gloomy and depressing situation for the migrants.

Migrants from the other part of subcontinent carried very few belongings, which were "expensive" as those were few. All around the Indian subcontinent, communities coexisting in that part of the region approximately a millennium assaulted each other causing sectarian violence, with Hindus and Sikhs on one side and Muslims on the other. There was a mutual genocide as unexpected as it was unprecedented. Scars of the hard times appear so strongly in this poem implying anxiety and turbulence in the speaker's mind. A natural calamity, flood, has been used to signify the situation, which has become hard to consume, has been diverted to another direction (Kamal 1997, 48). Sense of loss is intensified in a situation when people have to relinquish their belongings Next selected poem "Rain on Moss" has a suggestive title reflecting blessing of rain on moss. From ecological perspective, moss helps in soil erosion by absorbing heavy rain water. It equally helps in controlling floods and makes the soil fertile. The title also implies colonization as strengthening the sense of belonging. The poem is suggestive of future stability and prosperity. However, image of "water" appears as lethargic, bizarre, and idle rather than source of life. Water moves with a dull pace in this poem unlike floods, which illustrates that time significantly affects the conditions in postcolonial landscape. Though water runs with a slow pace, it carries an exaltation signifying joy and pride as if it has achieved something. "Bruised" is an image for the past blemishes. Rain may be symbolic of injustice. "Partial congruence" shows the confusion of the situation after decolonization constructing a paradoxical relationship between postcolonial subjects and memory. It is significant in the process of making individual as well as collective identity, which is "bruised," however, "exultant" (Kamal 1997, 48). Kamal contemplates on nature that arises in him desire and longing for the glory of the world to be restored. Here image of the river is "broad" yet it welcomes people to "embrace" (Kamal 1997, 48). Migrants thus find acceptability in new place.

The image is further extended to the portrayal of hard conditions of the world around him. "Cliff" symbolizes encounter with troubles and hardships ahead with frail "footholds." However, it equally portrays vigor and energy of life resulting from the hardships (Kamal 1997, 48). It depicts the ambivalence in the poet's mind suspended between thoughts of past and future. It also implies his sensitivity to the changing world around him. However, like an imagist, his poem lack rhyming pattern. This irregularity and fragmentation connotes celebration of memory portrayed in precise but vivid manner. This poem also reflects replacement of abstraction into concrete details transforming everything to solid dry images. Another insightful image in this poem is the willow tree. Willow is a fast growing tree with catkins is its part allowing the tree to reproduce. Such an image is to draw forth regeneration and fertility of life. Nature enlivens life. Willow catkins symbolize strength, stability, and firmness around society. Its movement as "swirling" denotes confusion of the migrants whose life resembles its course.

The poet has posed a question through an aesthetic image if happiness and prosperity can be hoped in the postcolonial world. "Matrix" that is the cultural, social, and political environment are ready to develop notions of opulence and prosperity. The poem foreshadows change. Image "matrix of joy" propels people to access and apply their personal powers for creating inspired models of reality based on joy, peace, security, compassion, overall well-being of humanity, and a greater sense of connection with one another.

"Anniversary" opens with a thoughtful contemplation on the situation of postcolonial world. Word "cascading" in the opening lines indicates the speaker's acquiescence to the challenges of revisiting memory lane and recollection of past painful experiences. However, their existence cannot be ignored since the wounds those memories have left are everlasting to keep his heart in anguish. The "terrain" left behind is not easy to forget (Kamal 1992, 123).

Keeping in view that Kamal's work had been published during postcolonial era, globalization, after World Wars, and the tragedy and disorder that intruded simultaneously, his poem "Anniversary" discusses how those events created a negative impact on nature and ecology of Kamal's "terrain." The opening word of the poem "cascading" also signifies the flowing and gushing effect of water over the land, the "terrain," which implies speaker's inclination toward nature and ecology evident through his lexical choice.

Kamal uses images from nature to satisfy his desire for belongings. "Chopped up moon," "wet stones" and "paddy field" refer to his nostalgia. Images of the pat are "floating signifiers, churning and flickering" (Ashcroft 2006, 2). Partition created "chaos." Emotional associations from their distant past strike the life of migrants in an absurd, unexplained, and recurrently agonizing and self-defeating manner (Kamal 1992, 123). "Chaos" usually refers to the property of a complex system whose behavior is volatile and which appear random, due to sensitivity toward small changes. It resembles the state of utter confusion, which keeps the mind perturbed. The situation mirrors the speaker's consciousness of the changes, which occurred after partition, as division of land often exasperates human beings. Individuals cannot be ripped off their origins; however, the profound reality is "imageless" (Kamal 1992, 123). In addition, images of "ashes and dust" in the last lines signify the celebration of belongings. Colonial aggression caused damage to the beauty and splendor of the land. Kamal reestablishes the status of a poet in reconstructing history. Like "a word-smith," he plays with words, well aware of the power and influence of language. Land that the migrants departed from remains in their memory as "an ancient ruin" and like "the sound of clog," it knocks the minds never letting it disappear (Kamal 1992, 123). Kamal reconstructs a unique sense of partition in this poem. He talks about the inseparability of land. Consideration of nature as something significant is important in disclosing history and the present.

Human sufferings and anxiety in the milieu of partition are endless causing disillusionment to their lives. It follows the idea that "inscription and transmission" destroy as well as create "transformation of personal memory into history and monument" (Saint 2010, 32). Language constructs a bond between nature and political history. Therefore, language and discourse are necessary tools to comprehend the political and social worlds that Kamal has described. Furthermore, ecocriticism and postcolonialism reconcile to establish an understanding of the natural, social, and political world in his works. Affiliations to place in ecocriticism and displacement in postcolonialism are consequences of colonization (Buell, et. al., 421–422).

Dislocation is not a failure but an expansion of cultural and aesthetic experiences of those dispersed (Mukherjee 1988, 3). Experiences of displacement and alienation generate themes of postcolonial literature. As a haunting figure, "the time lag" points out the return of the colonial repressed (O'Riley 2007, 1). The deployment of haunting suggests a desire to focus a theory of resistance. Use of symbols from ecology reflects Kamal's dissatisfaction over the disorder in the environment and his helplessness for not being able to find the truth with confidence of what is going on. The situation foreshadows dark future for the environment expecting "ashes and dust." Even the birds are disconcerted by the fragmentary changes in ecology. The poet is conscious of the damages colonialism has caused to the land as well as the living beings on it.

Appreciation of beauty alters altogether in the wake of colonial aggression. Conventional soft and serene imagery changes to crude and hard images. It is a source to recall exploitation and abuse. Nature imagery in "Anniversary" constructs in readers' mind an image of the world, which natural resources have turned blatant as a result of foreign violence. Kamal's poem "Kingfisher" displays his emotions and consciousness of identity. He relates images of nature to the individual's situation in postcolonial world, which has disintegrated and undermined the real spirit of identity. Nature images in this poem construct a fierce portrayal in reader's mind. It introduces readers to a sensory experience appealing to their cognitive senses of touch and feel. Drawing ruthless parallel between love and the kingfisher's beak implies the writer's attitude to emotions in the milieu of colonial past. April stabs like the kingfisher's beak (Kamal 1992, 125). The poet might have chosen the month of April as it is the start of summer in the warm region when emotions are disguised under intense hot weather after spring season is over. Probably the poet attempts to provoke intense feelings prophesying anguish in the future.

Physical anguish overpowers psychological strains or aggravates it otherwise. "River's glad torment" paradoxically replicates the perplexity of situation in the context of partition as people are in turmoil whether to celebrate or mourn while leaving the land of forefathers behind. It also emphasis that colonizers left us a legacy to yearn and remain confounded. In other words, where we stand and how we go is history. Historical space is filled with nostalgia, the gray times, and the replication of facts. This leads one to discover many aspects of life and settlement. The situation of looking at the past is "adventure, risk, transformation" (Mukherjee 1988, 3). "River" is thus personified and represented as a symbol of resistance equally suffering from "torment" the way people are chaotic and anguished. Maritime affairs are reconfigured by use of water and river images. A key observation is Kamal's repetition of combining the ideas of love and nature, which are also reflected in his poem "Anniversary."

Using image of "cloud," the speaker's mind advocates that if living together becomes intolerable, safe detachment should be preferred. Cultural clash in Indo-Pak subcontinent provided reasons for partition. Even nature does not allow unity with "disconsolate" objects. However, "clouds" separated from trees remain desolate and discomforted. Parting from origin leaves one in apprehension and ambivalence. Kamal's poem embeds the grief people suffered at the time of partition. Their future is shaped on a different "shore" (Kamal 1992, 125).

It elaborates Kamal's anticipation of future determined for people in some other region as a result of migration. At times, Kamal uses surreal images channelizing the unconsciousness as an instrument to unfold the power of imagination. Images from nature in "Kingfisher," for example, "tongue of water," and "cradle" portray the superiority of imagination (Kamal 1992, 125). Nature images also denote identity crisis. As always, stars show the directions and tell people their worth (Kamal 1992, 125).

Another poem "Winter Rain" reflects nature fabricating memories. Scattered nature images in the start, for example, "mist" to "rock-crystal's," jumping from "forest" to "moon" finally sketching memory in the form of "snow-fed stream." The poet connects distinct spaces of pre- and post-partition as ambiguous terrain between history and memory occupied by nostalgia, as a "no-go area ... viewed with suspicion and mistrust" (Sandru 2011, 220). Arising out of psychological, physical, cultural, and spiritual displacement of the imperialism era, the poet encounters experiences of longing, which are elevated by migration. "The natural environment is always a shaping force of individual and group psychology and identity" (Armbruster and Wallace 2001, 7). Problematic and chaotic in many ways, question of identity and loss compels postcolonial writers to consider themselves as historical beings. Kamal has attributed it an aesthetic feature defined as "the aestheticization of memory" (Poon 2016, 185). His poems represent "self-conscious musing" by representing memory in nature images. He does so by deliberately intermingling various nature images, ethno-culture issues, colonization, and the after effects. Juxtaposing words such as "carnage" and "rose-valleys" attempt to show purity of lands affected by foreign aggression causing bloodshed (Kamal 1992, 122).

The attempt to "aesthesize memory" is understood affirming the "the plurality" of poems in the backdrop of postcolonialism with words as "first light" and "wounds." It makes memory "artistic object so precious" and capable of encountering "the political function of memory" (Poon 2016, 185). Contradictions such as light and wound echo association to past as well as foreshadow hope. Aesthetics are purified during the process of sufferings. In Kamal's poems, a consistent discomfort with nature occurs when poet uses notions of aesthetics as refuge from the material world. Such an exercise of detachment is attempt to show how works of art represent themselves as beautiful and perfectly melodies knowing no boundaries of culture and geography "without any interest, goal or end" (Boehmer 2010, 172).

Kamal's poetry frequently emphasizes fierce aspect of nature at work, which establishes nature directly influencing human lives. Sharp imagery in the backdrop of postcolonialism reflects how colonial aggression might have caused destruction to the land. His poetry pays veneration to the objects of beauty in a postcolonial landscape. Nevertheless, his perception of the natural objects is quite brutal and wild, which reminds the readers of colonial aggression in pre-partition India. His works show resistance of postcolonial writer to the empire and its remnants by crude and unconventional images from nature. This lends his works an aesthetic sensibility, an eco-resistance. The relationship of objects of nature and human beings appears to be the dominant theme in his works, which shapes identity of the individual as well as the community. His works reflect how objects of nature have turned fierce and brutal in the milieu of colonial occupation. Even after decolonization, the land mourns and yearns for its liberty it once enjoyed.

Kamal's works contemplate the situation after partition and the environmental issues. He draws analogy between the two to figure out damages caused to both in the process of decolonization. Nature in his poems appears to lament physical and psychological conditions of individuals even after liberation. It is obvious that natural objects have turned fierce and have lost their tranquillity and composure the way human beings have done so.

Conscious of dependence on imperial powers, Kamal wishes for the recognition of a collective identity and consciousness by showing affiliation to nature and environment in his works. Such affiliation to the natural aesthetics features political movement and partition of Indo-Pak subcontinent in 1947. His works portray sense of place and nature geographically rooted into colonial aggression establishing its connection to environment. Unlike traditional poets, portraying glowing images of natural world, he demonstrates resistance, destruction, and suffering. Hence, images like "disconcerted trees, river's glad torment, startled dreams, blind earth, ancient ruin, chopped-up moon and ashes and dust" represent struggle against the colonial occupation and the repercussions.

Natural environment is regularly engaged in Kamal's poetry as a strong means to shape resistance. His approach to nature and postcolonialism show how literature discusses varying realities. It further highlights the idea of postcolonial aesthetics, which emphasizes form, taste, and experience as valuable areas depicting power relations around the globe. His imaginary world fashioned a unique strategy of nationalism in the newly born state as a result of partition. He appears engaged with pre-partition thoughts as well as the new environment if it could be called "new." He attempts to create a balance between the two searching for *one* identity. His works reflect his personal identity, social relationships, and "aesthetics sensibility" (Zuzana 2006, 108).

CONCLUSION

Environment, aesthetics, belonging, and postcolonialism integrated in one united whole is the proposed idea of this study to understand literature in the contemporary world. It enhances the scope of literature and illuminates its practical role. To conclude, "passion for land where one lives is a start; an action we must endlessly risk" (Glissant 1997, 151). To endure "risk" in the sense as conflicts over land teach lifelong lessons. Such devotion to land attributes "exceptionalism" as well as "difference" (Glissant 1997, 148). Until nature does not come across resistance, individuals will find no supporters. Kamal's aesthetics of belonging resist conventional mysticism found in natural objects. His appreciation comes in the form of acrimony hence enlightening "the political force of ecology" (Glissant 1997, 151). Kamal talks about "the aesthetics" of his land, "the blind earth, the empty horizon, paddy-fields, ashes and dust, snow-fed stream, moss-grown stepping-stones." He questions ecological challenges whether such poor condition of environment could be a source of contentment. Like Spivak, he welcomes "inexhaustible taxonomy" and for him, this despondent condition is not a weakness but source of solace and serenity generated out of "aesthetics of turbulence" (Glissant 1997, 151). It shows his concern to maintain identity and belonging. His works also reflect a resistance to stop environmental degradation. It is a unique way of imagination to consider fragmentation during the process of decolonization. Fragmentation in Kamal's poems replicates fractured memory. His strategy is to preserve identity. What is lost cannot be restored; however, at the same time, it cannot be diminished altogether from one's memory and remains source of regaining energy and spirit. Thus, "tempering of post colonial environmental imagination" becomes his strength (DeLoughrey 2011, 29). His works are significant in pointing out a need for appreciation of humanity particularly people of his land carrying scars of partition facing environmental challenges. His nature imagery exposes the readers to new experiences creating an entire world. He moves the readers across a postcolonial landscape echoing conventionality sometimes celebrating memory and showing passion to belongings.

The moon thaws
Before your loveliness:
You are the breathe of violets. (Kamal 1992, 122)

BIBLIOGRAPHY

A Brief Guide to Imagism. 2017. Poets.org. https://poets.org/text/brief-guide-imagism.
Armbruster, Karla, and Wallace, Kathleen. 2001. *Beyond Nature Writing: Expanding the Boundaries of Ecocriticism.* Virginia: University of Virginia Press.
Ashcroft, Bill, et al., ed. 2006. *The Post-Colonial Studies Reader.* London: Routledge.
Berry, Ashok, and Patricia Murray, eds. 2000. *Comparing Postcolonial Literatures Dislocations.* New York: Palgrave.
Buell, Lawrence, et al. 2011. "Literature and Environment." *Annual Review of Environment and Resources* 36: 417–440. http:/environment.harvard.edu.
DeLoughrey, Elizabeth, and Georg Handley. 2011. *Postcolonial Ecology.* Oxford: Oxford University Press, Inc.
Gandhi, Leela. 1998. *Postcolonial Theory: A Critical Introduction.* New Delhi: Oxford University Press.
Glissant, Édouard. 1997. *Poetics of Relation.* Michigan: University of Michigan Press. DOI: 10.3998/mpub.10257.
Kamal, Daud. 1992. *Rivermist.* Peshawar: National Book Foundation.
———. 1997. *A Selection of Verses.* Karachi: Oxford University Press.
Mardorossian, Carine. 2013. "Poetics of Landscape: Édouard Glissant's Creolized Ecologies." *Callaloo* 36, no. 4: 983–994. DOI: 10.1353/cal.2013.0196. 2013.
Moslund, Sten. 2015. "Postcolonial Aesthetics and the Politics of the Sensible." In *Literature's Sensuous Geographies. Geocriticism and Spatial Literary Studies.* New York: Palgrave Macmillan.
Mukherjee, Bharati. 1988. "Immigrant Writing." *New York Times Book Review.* https://www.nytimes.com/1988/08/28/books/immigrant-writing-give-us-your-maximalists.html.
Nayar, Pramod. 2008. *Postcolonial Literatures: An Introduction.* New Delhi: Pearson Longman.
O'Riley, Michael. 2007. "Postcolonial Haunting: Anxiety, Affect, and the Situated Encounter." http://postcolonial.org/index.php/pct/article/viewArticle/728.

Poon, Angelia. 2016. *Transcultural Aesthetics and Postcolonial Memory: The Practices and Politics of Remembering in Tan TwanEng's The Garden of Evening Mists*. Cambridge: Cambridge University Press. DOI: 10.1017/pli.2016.4.

Saint, Tarun. 2010. *Witnessing Partition: Memory, History, Fiction*. London: Routledge.

Sandru, Cristina. 2011. "Postcolonial Nostalgias: Writing, Representation and Memory." *Journal of Postcolonial Writing* 48, no. 2: 201. DOI: 10.1080/17449855.2011.639937.

Su, John J. 2011. "Amitav Ghosh and the Aesthetics Turn in Postcolonial Studies." *Journal of Modern Literature* 34, no. 3 (Spring): 65–86. https://www.jstor.org/stable/10.2979/jmodelite.34.3.65.

Wood, Lawrence. 2015. "The Environmental Impacts of Colonialism." http://vc.bridgew.edu/honors_proj/119.

Zuzana, Vasko. 2006. *Aesthetics, Authenticity, and Belonging: The Roles of Personal Identity and Aesthetic Sensibility in Artistic Development and Practice*. Burnaby, BC: Simon Fraser University.

Chapter 7

I Am a Tree Leaning

Neocolonialism, Eco-consciousness, and the Decolonized Self in Margaret Atwood's Surfacing

Anik Sarkar

Canada had its share of colonial struggle, first as a French colony in the sixteenth century and then as a colony of Britain in the eighteenth century. It only tasted freedom, just as most of the colonized nation-states in the first half of the twentieth century. The aboriginals of Canada were put through violence, displacement, trauma, and poverty, not to mention diseases like smallpox, measles, and yellow fever that were nonexistent in its lands, were later introduced by the colonizers. The small pox depopulated a large section of the natives (Cadotte 2015). Colonization—an organ of imperialism—fosters an imbalance of power relations, while being a breeding ground for dehumanization, relentless exploitation, plundering, and hostility that leaves people disconnected to their lands, culture, identity, language, and environment. Since Enlightenment rationalized the search for "truth" as a sign of civility and progress, colonizers formulated a moral excuse to embark on the task of dispelling darkness from "uncivilized" cultures that mainly belonged to the continents of Asia, Africa, and places of aboriginal dwelling in Australia and America. The rich mineral deposits in these places along with spices, rubber, timber, fur, cash-crops, and other products that were native to the indigenous people were ferried to Europe, which flourished economically (Murphy 2009). At a time, the colonization process sidelined with the Industrial age in Europe and there began an intensive exploitation of environment and natural resources available in the colonies, which in the process were invariably left tattered and deprived.

Capitalism, having its roots connected to colonialism, prompts the destructive exploitation of the environment partly through its reluctance of using eco-friendly technology, because using them otherwise would add up to the manufacturing costs and impact one of its core "tactics" of generating maximum profit at minimum expenditure. Throughout the past century, the absence of eco-friendly technology, greed and reluctance in fostering an ecological empathy lead on to hazardous implications as ice caps melted in the polar regions, global temperatures increased, climate changed, ozone layer perforated, toxic pollutants released in the air, rivers polluted, species extinct, and to the recent uncontrolled bushfires that wreaked havoc in Amazon and Australia and the declaration of a climate emergency; such has been the condition of the earth. Postcolonial environmentalism seeks to unravel the effects of colonialism on the environment and offer a counter-discourse. The ideas that emerge from such projects, allow us to scrutinize the deeper effects of colonialism, widening our perspectives on the impact and highlighting the oppressive connections that otherwise evade a general summation. Atwood's novel *Surfacing* tracks the impact of colonization as a disease that lingers on, years after the departure of colonizers, by advocating a suitable ground for neocolonialism to begin, which preserve and emanate the essence of imperialism. Neocolonizers influence and control the politics, culture, and economy of countries through capitalism and cultural imperialism, which is disguised in the form of "globalization." I look at three aspects in the novel, namely, neocolonialism—which Atwood presents as "Americanization," eco-consciousness—where the narrator-protagonist comes to a realization of multifarious violations that are exercised on her, around her and her environment and decolonization—which is her way of responding to these violations at a personal level.

Language is one of the fundamental areas that is first affected by colonialism. The "colonial process" has its inception in language as language creates reality (Ashcroft et al. 1995). We ascribe meaning to the world through language and we communicate our feelings, emotions, and experiences through language. Very often, Canada as a nation is portrayed to be linguistically divided between two groups, an English-speaking population and French-Canadians. As *Surfacing* begins, we come across an unnamed narrator traveling to her place of origin, that is, rural Quebec with her boyfriend Joe, and two of her friends, Anna and David. The narrator speaks English and had taken lessons in French when she was in school, but she is not affluent in the language. The complexities borne out of the divisions in language groups, is first encountered in the novel through the presence of multiple signboards, as I quote one of them below, described by the narrator, "THÉ SALADA, BLUE MOON COTTAGES ½ MILE, QUÉBEC LIBRE, FUCK YOU, BUVEZ COCA COLA GLACÉ, JESUS SAVES, mélange of demands and languages, an x-ray of it would be the district's entire history" (Atwood 2012, 13).

The mixing of multiple languages would imply a heteroglossic space of voices, desires, aspirations, cultures, and a celebration of differences although we later discover that this fosters an exercise in hostility rather than cooperation. The narrator too feels distant from her place of origin, because of her inability to properly translate the words in French. Also, throughout the novel, we find that there had been a rising tension between all of the separate language groups. Even, the English-speaking Americans had their distinct differences from the English speakers of Canada, while few Canadians were gradually getting "Americanized" in an unconscious "mimicry." In rural Quebec, the narrator notices a woman wearing slacks, which was prohibited earlier by an old priest. She remarks on how things had changed since then, with people mocking her French in a place that no longer has English speakers:

> "Amburger, oh yes, we have lots. How much?" she asks, adding the final H carelessly to show she can if she feels like it. This is border country. "A pound, no two pounds," I say, blushing even more because I've been so easily discovered, they're making fun of me and I have no way of letting them know I share the joke. (Atwood 2012, 27–28)

If there are more than one language groups, formed out of the colonial experience, there is always a tension that is shared among these groups.

Another way of looking at it would be to assume that a dominant language group in a particular area, such as rural Quebec, suppresses other minor groups, and all the other groups are expected to learn the language of the majority. Failing to do so often results in humiliation and alienation, be it self-induced, as in case of the narrator who was clearly embarrassed at her incapability of speaking proper French, and also by the members of the dominant language group like the woman at the counter alongside the two men, who sneered and mocked the narrator for her attempts to conceal her identity. The narrator takes us back to another incident from her childhood, where her mother (an English speaker) and the wife of Paul, Madam, (a French speaker) were trying to have a conversation, but unable to make proper sense, past greetings. The differences rising out of language has been paid close attention to by Atwood, who uses it to mark out the crisis borne out of a colonial past. The original language of native Canadians is lost, and only the language left by British and the French remain as the main languages, and hence of a colonial origin. Even then, there are problems arising out of both these language groups, and likewise a superpower neighbor like America has a massive cultural influence, owing to the internal disparity and conflict that rises out of cultural and linguistic differences in the people's struggle for identity.

If the language problem in the novel that causes disparities can be attributed to a colonial cause, there are several complications arising out of the "newer" form of colonialism. Suman Makhaik writes:

> In Surfacing, issues of gender domination that is sexism, and environmental degradation that is naturism, are bound up with aspects of socio-political, economic and cultural imperialism, and the narrative structure of the novel reveals a "matrix of oppressions" in which dichotomies such as self/other, mind/body male/female, American/Canadian are correlated with the culture/nature paradigm. (Makhaik 2011, 129).

Indeed, in *Surfacing*, nostalgia is followed by anguish, as things replace things, materiality and consumerism grow, while the "nativeness" is replaced with an "alien" parlance. The binaries are clearly brought out against one another and they reveal startling imbalances and as the splits widen, we are aware of the deep cuts and "amputations," which haunt the narrator. At one level, she feels a separation from nature as the toxicity of modern culture pervades throughout rural Quebec, while her own separation from her child, her father, and her former lover appears as other separations in her unconscious, which germinate a desire to separate from masculine culture by embracing nature wholeheartedly in order to synthesize and become one with it. This desire to synthesize with nature and become plant-like echoes Atwood's other novel, *The Edible Woman* and Han Kang's 2016 Man Booker International prize winning novel, *The Vegetarian*: "The protagonists of *The Vegetarian* and *The Edible Woman* are young women who feel trapped and constricted by society's strong patriarchal conventions, attempting to escape them by eschewing meat, equated with the exploitation of women, animals and the environment" (Ferreira 2019, 147). "Becoming tree" is a resistance offered to the unequal power hierarchies that dictate behavior, norms, and etiquettes. It is also a resistance to corruptions, complexities, and anxieties of modern life. Sumana Roy in her nonfiction book *How I Became a Tree* mentions that escaping noise is an urgent need for becoming a tree—one being the noise of humans and another reason would be to embrace the silence of trees (Roy 2017). Later, in the novel, the narrator's utterance "I am a tree leaning" adheres to her "becoming a tree" by leaning against one, and signifies the act of metamorphosis that merges the contraries—her split body and mind.

The colonial and neocolonial impact leaves people disconnected to their culture, language, and origin. The foreign influences and practices establish grounds for a mixed, hybridized culture, which uproots people from their native identity and alienates them; and quite often they begin to "mimic" the dominant culture as they are separated from their former one—which they consider to be inferior. The narrator refers to her original home ground as "foreign territory," as it is far removed from recognition. Her father's memory is a constant reminder of her past, her identity and her connection to the older place of her childhood. As she notices the changes in the landscape, especially their favored road, she cannot believe that a new road had upturned their possibility of taking the older route. When the narrator and her friends reach their destined location earlier than her anticipation, it should have been a moment of joy but it only brings a sense of loss and deprivation. The shortcut had been achieved by blasting rocks and bulldozing trees. Advancement brings no comfort to the narrator, because it comes at a cost of exploitation. She feels that it if she would have arrived at the lake in a nauseated state because of the tiring journey, she would have felt better, "But they've cheated, we're here too soon and I feel deprived of something, as though I can't really get here unless I've suffered; as though the first view of the lake, which we can see now, blue and cool as redemption, should be through tears and a haze of vomit" (Atwood 2012, 13).

As she explores rural Quebec, she develops a constant dissatisfaction with transformation and change—the landscape of her former place, the habits of people that have been altered, things that shouldn't be there like the motel, and the presence of her friends (who seem to be disrupting her relation with her former attachment with the place)—speaks about her desire to return to a pre-neocolonial era, the time of her childhood. In her mind, there is a conscious simulation of the former place from her memory while her lived experience of that place is constantly jeopardized by undesirable foreign elements. Her struggle to accept the present condition and her attempt to recreate a space out of her fond memories can be compared to the agitating struggle of a nation to reformulate its authentic past experiences, before the arrival of the colonizers.

The precolonial landscape and untouched authentic culture acts as a touchstone for the narrator, while the changing contours, the borrowed habits, and coarse language are understood as a "decadence," an off-shoot of "Americanization"; a performance of "mimicry." The narrator is astonished at the presence of multiple stuffed moose at the gas station, dressed in human clothes, and portrayed to be imitating Americans: "a little boy moose in short pants, a striped jersey and a baseball cap, waving an American flag" (11). The only intent of placing these was to attract customers, and to get them to pay. The humanization of animals is as obvious as it is to say, a task of stripping off their "animality," of forced homogenization and the eradication of identity. To find, that the Americans had not only Americanized humans in Canada, but were setting up stuffed animal as American citizens, had unsettled the narrator. Her friends' visitation to her place leads her to believe that she was a representative of that place, and responsible for any decadence that would otherwise leave a negative impact on their minds. So, she says "defending" herself, "Those weren't here before," as though she was liable for the blatant capitalist tactics that her rural place had adopted (12).

The novel in its progress suggests first, the pre-Americanization of Canada as an expanse of verdure and purity, while years later in the unnamed narrator's visit, she finds white birches dying, disease spreading on a wide scale. The landscape of her former place, rural Quebec, is subjected to rigorous change at an unprecedented level, which leaves her in a state of shock and disgust. Her childhood was spent in a serene environment, among trees, clear lake, fresh air, and organic food. As she travels through the roads, she finds that the old roads have been replaced by newer ones, she remarks, "Nothing is the same" and "why is the road different, he shouldn't have allowed them to do it" (10). The reminiscence of a time long lost is a nostalgia craved by postcolonial writers as well as characters with a colonial past, desiring a reversal that would somehow restore the conquered lands, undoing the damage dealt in the process of capitalizing on native resources. Several areas of the forest were littered with trash, orange peels, tin cans, greasy paper, which the narrator remarks to be "tracks" left by humans. This incident is being compared to dogs marking the territory by the act of pissing. The narrator and her friends then come up to a dead heron, which had been brutally killed and hung from a tree. She assumes that the heron had been killed as it was of no other use: couldn't be eaten, did not have a commercial value like training it to talk, so destruction had been the viable way to demonstrate power over the helpless. The Americans were stuffing seaplanes with illegal fish, killing trout with dry ice and letting loons get slashed in the propeller blades of their powerboats.

Suman Makhaik also mentions in her work regarding Americanization as associated with "metal": "Metal symbolizes cold hardness and, therefore victimization. Americanism, destruction, and death become easily associated with metallic objects. They are shown to assault nature as well as individuals. Trees are cut with the help of metallic instruments; steel is used to clear the forests and to 'invade nature' " (Makhaik 2011, 141). Use of metal and metallic objects denotes rapid industrialization where the simple handmade tools derived from nature like stones and wood are discarded. As foreign industries and companies start to enter the regions of a developing and underdeveloped nation, the people in those areas are distanced from their connection to nature. They stop using the goods that can be derived from the environment considering it to be outdated, while using artificial products developed and mass-manufactured in the industries. This practice not only hampers local tradesmen but also affects the mentality of people who consider foreign goods to be "superior" and "advanced" than what they can locally produce. Slowly, they also begin to adapt to the foreign culture and lifestyle that is associated with consumption of foreign goods, and gradually feel alienated from their own roots. The narrator notices how everything is expensive in rural Quebec than in her city, animals were not being domesticated, and bread was available in "wax paper wrappers" (28). All these are signs of capitalism taking over these areas of Canada, which were supposed to be fertile and replete with natural resources. The Americans were buying properties to establish businesses, and people were not producing locally made goods, but importing from other production hubs.

The narrator is an artist for publishing houses, and as she draws for her latest project, a children's book titled *Quebec Folk Tales*, she remembers a particular incident where her publisher Mr. Percival commented that one of her drawings were "disturbing" (63). As she defended herself stating that children like to be frightened, Mr. Percival replied that it is the parents who buy the books. Further on, the narrator adds that the publisher wants to impress the foreign English and American publishers, demanding imitations that follow glossy market trends. Atwood's projection of this shift in target audience for a children's book depicts the stripping away from the publisher's core motive of imbibing knowledge and imagination into the minds of young readers, uncovering a deeper ideological concern for an investment. This investment targets maximum profit, as the narrator remarks that the color choices of the paintings have to be limited, to cut down publishing costs.

The impacts of neocolonialism and the destructive implications of capitalism are further extended in the novel through how the treatment of landscape and the environment has been scrutinized by the narrator. She remarks that the Americans catch more fish than they can eat, and sarcastically states that they would blow the fish with dynamite if they could get away with the killing. This statement implies that the narrator connects violation of nature to how the Americans had been carelessly handling the resources in Canada, as she further extends her disdain by bringing into focus an incident where she met two Americans on her way to the bass lake. One of them burned his boots on their campfire, while also unable to land his fishing lure on water. They carried advanced equipment like "automatic firelighters," "cook set with detachable handles," and "collapsible armchair," but they failed to capitalize on their task, which would have been possible with minimum tool kit; she sarcastically ends her recollection by stating: "They liked everything collapsible" (81). A machine-dependent life form often leads to the lack of skill set, tarnishing survival and physical abilities. This is noticed by the narrator when she thinks of David's and Joe's inability to cope up with a rural lifestyle in the forest, when they are cut off from habitation, "I measured their axework with my father's summarizing eye. In the city he would shake hands with them, estimating them shrewdly: could they handle an axe, what did they know about manure?" (101–102)

The chopping of the log was sporadic, uneven, and destructive. They filmed this event of cutting a log from the forest as achievement, which the narrator compares to the task of hunting a lion or rhinoceros, and posing with it as a trophy. David, one of the narrator's friends, is a startling exemplar of a "mimic man." David wanted to film their tour of rural Quebec and call the documentation as "Random Samples." The filming undertaken by them was a cultural import from America, as the narrator remarks that even though David and Joe couldn't afford a camera, they hired one in order to follow the trend. The film would not have any specific purpose, but would be clippings of random events, juxtaposed to make an incoherent assemblage of footage. This pseudo-activity of making a film for no specific reason or purpose, by compiling fragmented clips symbolizes the fragmented self of David, imitating and mixing bits and pieces of popular thoughts and utterances from here and there, as contextualized later in the chapter.

The connection that humans share with the nature world—the world of nonhumans, both living and nonliving, is an inter-dependent one. There is a complex network that functions in binding all the aspects of existence in this plane of reality. The realization of this symbiotic network and understanding how humans are a part and an active component of their surroundings is dawned upon few individuals who keenly observe, rationalize, and internalize their immediate reality. As we see in the novel, it required an acute empathy for the environment, some form of provocation or "suffering" and perhaps a critical distance to safely observe and reflect on these deeper levels of reality that function beyond socioeconomic and cultural constructions. After witnessing the rapid changes in her place of origin and contemplating on her own experiences of subjugation, the narrator becomes aware of the power structures that function in society. These structures that have been raised to violate nature and women can be understood through a conscious "awakening" as is clear in the case of the narrator who is deeply moved and hurt if she finds a disruptive force in action. Her becoming eco-conscious is a progression that develops gradually in the novel, and in this process, she does not consider herself to be someone who is separate from nature.

Since the beginning of the novel, the narrator displays a fond sympathy for her place of origin. On a quest to find her missing father, she unveils deeper problems that took shape since she left the place, problems on one hand, arising from her personal guilt and also problems that were on a collective level, affecting rural Quebec, which moving from the particular to the universal can be signified as the natural world. This is because, the particular woman as well as the name of the exact place, which she belongs to, has been unnamed. In the context of this novel, one reason for withholding the naming of an individual or the place of her origin could be a viable choice if the author seeks to universalize the particular. There are several instances in *Surfacing*, where the feeling of the unnamed narrator is the general feeling of women who have been marginalized. Similarly, the environmental violations carried out in the name of development and profit were not just particular to rural Quebec but a global phenomenon. The atrocities and rigorous exploitation carried out on women is hence often paralleled and compared to the exploitation of the environment in many ecofeminist and postcolonial novels. Careless treatment of the environment, senseless killing, and demonstration of brutal force are thus a demonstration of capitalism/toxic masculinity in action. The act of senseless killing in the novel is a characteristic of the "machoism" of David, as he remarks that he wanted to hook a split beaver, which, according to him, should have been the emblem of Canada. "Split beaver" explains David is a joke for a female sexual organ, to which the narrator reflects about the absurdity of equating parts of human body to dead animals.

David is a staunch patriarch, an exploiter of women and his acts are very similar to the exploitation of nature carried out by neocolonizers who time and again have been portrayed either directly or indirectly in the novel. Though he is vocal about the American businessmen, calling them as "Yank pigs," elaborating on "industrialization" and "exploitation," he does not realize that he too aligns with the things he critiques by exploiting the emotional needs of woman, dominating them, reducing them to objects, and ridiculing them (123). Anna is humiliated often, and used as an object to fulfill the whims and fancies of David. This subjugation is experimented on the narrator who denies the crude approaches of David. He constantly uses lewd sexual remarks toward Anna and the narrator that involves the subjectification of women and their bodies.

The silence of the narrator to such remarks in the beginning could explain the deep-rooted patriarchy functioning in society, which has been accepted by men as well as women, "normalizing" the reduction of women to objects. David's refusal to see Anna without her makeup suggests the hollow adherence of people toward superficiality. Anna complains to the narrator that David has been unfaithful to her, having encounters with multiple women while narrating about them to her later. David uses his masculinity to forge a power over Anna and carry out encounters outside of marriage while having "theoretical" excuses to defend his actions. The othering of Anna is rampant and this othering extends to not just humans, but nonhumans too, as we see the heartlessness of David when he comments on beavers and heron. David took things to the extreme when he forced Anna to shed her clothes for a video and humiliated her in front of everyone. He made sexual advances toward the narrator, and lied to Anna that he had slept with her. The narrator becomes the victim in the group, being the "purest," which contradictorily becomes a "heresy" as everyone else had been unfaithful to their partners. When David accuses her of hating men or having the desire to turn into men, the narrator starts to think if it was true that she hated men. She then goes on to feel that the entire incident is an impact of Americanization, which strengthens her disdain toward modern "culture" because everyone around her had been symptomatic of apathy and degeneracy.

As the narrator goes through some old scrapbooks, she scans her brother's childhood drawings that were filled with images of violence and bloodshed. Most of the drawings probably inspired by comic books speak about how the imagination of children, especially boys had been impacted by ideas of war, absurdity, evil, monsters, and death. Whereas, the drawings of the narrator consisted of "ornately-decorated Easter eggs," rabbits, flowers, grass, trees that looked "normal and green," symbolizing that the girl children had their imagination working innocently and with a keen ecological sense even if it was unconscious in that tender age (115). The narrator avoided social gatherings, since she was young. She despised birthday parties and getting social with people, habits that had grown on her. She noticed how she was often targeted by other children, for her being distant and reserved. She was often bullied as she recounts being tied to fence, gates, and "convenient" trees. She reflects on these episodes: "Being socially retarded is like being mentally retarded, it arouses in others disgust and pity and the desire to torment and reform" (88–89).

Perhaps this is why the narrator characterizes a "distance" with her surrounding including her two friends and Joe, even though the "withdrawal-to-self" does not maim her or make her incapable to attend to other's needs, as she actively cooks meals, guides them through the island, and helps them fish. She is "absent-present" where she is supposed to socialize, and she is hesitant to participate in many events. She rarely talks to her boyfriend and abstains from deep emotional attachment claiming that David and she are alike, unable to love, while Joe and Anna are the sufferers. Although, this form of "distancing" allows her a critical gaze, exercising, which she is able to continuously observe, weigh, and analyze events in her lived experience, sometimes also retreating to her past, facilitating a comparative take on things.

She also feels the rejuvenation of the natural environment around her, as in the city, she feels sick of the effects of air-based pollution that makes it hard to breathe, the harsh heat and pollutants that stick to the skin. The environment is adversely affected in the cities due to the growth in pollution of various kinds. That makes living in the urban spaces much more difficult than in villages or small towns in the hills that have a good coverage of trees. The narrator's disappoint comes from the fact that the place of her childhood is getting robbed off the safety and comfort of breathing easy. Rapid urbanization is taking over the rural areas, which are covered with trees and water-bodies, only to be converted into tourist spots or production centers. Huge number of trees had already been felled, many areas in the lake had been over-fished, the lake had been altered to fit the need of the capitalists and the entire area was being transformed. A reservoir was being planned at the spot, while an offer had been made to the narrator for selling off her parental cabin, which would be turned into a resort. Her awakening to the environmental problems in and around rural Quebec that also paralleled the exploitation of women made her conscious about deeper problems and how their sources were interconnected.

The awareness of "coloniality"—that it not only colonizes the body but the mind too—is a step toward decolonization. Without this awareness, it is impossible to decolonize oneself, as just getting rid of the colonizer does not secure a complete decolonization. One's body can be free in a province that no longer has a colonizer, but the mind is still vulnerable as habits, routines, cultural impacts, and the ideologies of colonialism are deep-rooted in its recesses, and it is the awakening up to this actuality, which can initiate the process of decolonization of the mind.

The search for her father led the narrator to believe that he had gone insane living in isolation, cut off from society. She finds some paintings and sketches that at first do not make any sense. Later, she discovers that they actually represented a location under the lake, where aboriginal paintings were supposed to be found submerged in the deep. Taking a dive, she starts to look for the paintings, according to her father's demarcations. But in the depth of the waters, she is gripped with fear, as the image of her aborted child floats below her like a shadow. Her traumatic episode at the abortion clinic had been deeply embedded in her unconscious, which now manifests in the lake as an image of guilt. She remarks that "they had planted death in me like a seed" (184). This psychotic breakdown brings about a significant change in her. She relates the event to her father's, as well as her own vision of truth, wherein her father had discovered through his hobbies, the zone of salvation of the Indians—the sacred place where truth could be learned. She is filled with an enormous sense of guilt, for having to abort her child and perhaps she too aligned with the patriarchal order, the destructive force of colonialism that plunders the environment.

Her undoing of this alignment dawned a startling change of perspectives: "In the cool green among the trees, new trees and stumps, the stumps with charcoal crusts on them, scabby and crippled, survivors of an old disaster" (191). She becomes a survivor of the harm brought about by the excesses of modernization and modern life: consumerism, exploitation, and neglect. She does not want to live a shallow life anymore, under the guise of civilization but wants to return to the authentic experiences of being in the world, in the natural order, sharing the visions and experiences of other nonhuman beings, "Sight flowing ahead of me over the ground, eyes filtering the shapes, the names of things fading but their forms and uses remaining, the animals learned what to eat without nouns. Six leaves, three leaves, the root of this is crisp. White stems curved like question marks, fish-colored in the dim light, corpse plants, inedible" (191).

Further on, she thinks about the possibility of being a filament plant with hair sprouting out. She reasons the purpose of the coffin as being a preserver of bodies, which wouldn't allow the transformation of them into something else within the earth. She identifies herself as a part of the earth, and she wants to cut off any other obstructions that otherwise would hinder her connection to the earth.

She negates her guilt of abortion by copulating with Joe in an open environment, discarding any barriers that might hinder the "purity" of birth. She wants the newborn to be divorced from the materialistic grasp of the world and states that the process of giving birth will be animal-like as the child "will slip out as an egg, a kitten" into a pile of dry leaves, and she will "never teach it any words" (209). She wants to counter her guilt of abortion through giving birth to a child in the most naturalistic manner possible, and also keep the child away from becoming engrossed in modern culture as she was aware as well as a witness of the destructive forces of "modernization" and "advancement." This new vision also allows her to see through people, as she sees through David, "The power flowed into my eyes, I could see into him, he was an imposter, a pastiche, layers of political handbills, pages from magazines, affiches, verbs and nouns glued on to him and shredding away, the original surface littered with fragments and tatters" (194–195).

In her description of David, it is clear that she is tired of the pretentiousness that people like him and Anna possess. She calls David a "Second-hand American," as someone who requires healing to be able to reach his true self (195). Her act of resistance was in destroying the film that had recorded Anna in a humiliating light—an act of voicing out for the subaltern. Further on, she completely takes on a frenzy, a return to wilderness, by sabotaging every object or act that made her feel like a "human."

While retreating to wilderness, and then emerging with a profound sensibility, she involves herself in an activism that aims at restarting and redoing the process of "becoming" human. This form of resistance that may seem "barbaric" and "absurd" was essential to decolonize her "self" from the ills of neocolonialism, patriarchy, and the modern society. She says, "From any rational point of view, I am absurd; but there are no longer any rational points of view" (218–219). This can be understood as, her being free from the gaze, be it imperial or patriarchal, which puts constraints and limitations on rights, freedom, and identity formation, and also contributes to the fact that these limitations are artificial constructs that disappear when their proprietors are absent. The animal-like state gives her an identity, courage, and power, as she says: "This above all, to refuse to be a victim" (248). It also offers her a motivation for countering Americanization: "They exist, they're advancing, they must be dealt with, but possibly they can be watched and predicted and stopped without being copied" (246).

The novel ends in suspense as the readers are left on their own to decide, if the narrator returns to civilization as Joe waits for her or does she change her mind in the last moment. This ambivalence is also a condition of the contemporary awakened human, who in spite of having an acute knowledge of the violations, power structures, and subjugation may find it difficult to counter the interconnected forces of oppression. Perhaps what the condition demands is the need for a collective awakening and activism toward marginalization and our environment.

BIBLIOGRAPHY

Ashcroft, Bill, Gareth Griffiths, and Helen Tiffin. 1995. *The Post-Colonial Studies Reader*. London; New York: Routledge.

Atwood, Margaret. 2012. *Surfacing*. Kindle Edition. UK: Hachette Digital.

Cadotte, Marcel. 2015. "Epidemics in Canada." In *The Canadian Encyclopedia*. Historica Canada. Accessed April 30, 2020. https://www.thecanadianencyclopedia.ca/en/article/epidemic.

Ferreira, Aline. 2019. "The Gendered Politics of Meat: Becoming Tree in Kang's the Vegetarian, Atwood's the Edible Woman and Ozeki's My Year of Meats." In *Utopian Foodways: Critical Essays*, edited by Teresa Botelho, Miguel Ramalhete Gomes, and José Eduardo Reis. Porto: University of Porto Press, pp. 147–160.

Makhaik, Suman. 2011. "Surfacing." *Ecofeminism in Margaret Atwood: A Study of Selected Novels and Short Stories*. (PhD diss., Himachal Pradesh University). http://hdl.handle.net/10603/121172.

Murphy, J. 2009. "Environment and Imperialism: Why Colonialism Still Matters." *SRI Papers*, No. 20. Sustainability Research Institute, University of Leeds, UK.

Roy, Sumana. 2017. *How I Became a Tree*. New Delhi: Aleph.

Chapter 8

For Appearances Must Deceive

Misreading the Environment in *Days and Nights in the Forest* and Its Cinematic Adaptation

Chinmaya Lal Thakur

The term "environment" in the title of this chapter is supposed to be broader and more radical than its Oxford English Dictionary connotation, that is, "the surroundings or conditions in which a person, animal, or plant lives or operates" (*Lexico*, n.d.). It refers—in the specific context of Sunil Gangopadhyay's 1968 novel *Araneyer Dinratri* that has been translated into English as *Days and Nights in the Forest* and its eponymous film adaptation by Satyajit Ray in 1970—to the way humans are usually disposed to engage with the environment that surrounds them and constitutes them as social beings. These surroundings include the forests outside Calcutta (now Kolkata) to which the four young men travel as they seek a break from the humdrum of urban existence. The Indian government officials who look after the Singhbhum area and its Santal and Oraon inhabitants and the members of the Tripathi household that they meet during the sojourn further comprise the whole environmental milieu.

Common to Gangopadhyay's novel and Ray's film is the inability of the four men to understand and take control of what they witness in the forest. Their experiences during the short excursion turn out to be very different from what they had expected. And, it is this significant discrepancy between what they had anticipated and what they actually see and undergo that forms the subject of discussion in the present chapter. The chapter discusses the cultural, social, political, and economic implications of the incongruity as being reflective of Gangopadhyay and Ray's critique of the usual way in which humans interact with the environment, that is, by assuming that they are the latter's masters and have the same under their control.

The suggestion about the relationship between humans and the environment as providing important insight into sociopolitical, cultural, and economic issues is drawn from Robert Pogue Harrison's magisterial study *Forests: The Shadow of Civilization* (1993). In this work, Harrison is not so much interested in providing empirical details about how human civilization has encroached upon, managed, and exploited forests as in underlining the role that forests have played in the cultural imaginings of the West. Analyzing writings by Vico, Dante, Shakespeare, Rousseau, Wordsworth, Conrad, Sartre, and Thoreau among others, he reaches the conclusion that forests represent the literal and imaginative limits of human civilization. And, he argues that the currently prevalent anxiety about disappearing forest lands is actually human insecurity about being unable to perceive boundaries to the way we make a place for ourselves in the world, to the way we clear the ground to make a dwelling for ourselves on the earth. It thus seems to us that without the "outside" constituted by the forests, it would simply be impossible for us to dwell within the "inside" (Harrison 1993, 247).

In the Indian context, Upinder Singh's expansive study *Political Violence in Ancient India* (2017) provides a detailed survey of Indian cultural imaginings of the wilderness in its eponymous penultimate chapter. Through her readings of epics like *The Ramayana* and *The Mahabharata*, treatises such as *Arthashastra* and *Nitisara*, and didactic literature comprising *Jataka* and *Panchatantra*, Singh suggests that there wasn't a unique or singular imagination of the wilderness in ancient India. The state, nonetheless, continued to interact with the forest and its inhabitants and the relationship was marked with conflict, difference, interdependence, and incorporation (Singh 2017, 457–459). Sadly, a similar analysis of literary and cultural representations about the human and forest relationship in colonial and postcolonial India doesn't seem to have been undertaken as yet. The present chapter, in that Gangopadhyay and Ray's temporal context is of the early decades following India's independence in 1947, would have certainly benefited from such a work.

Ashim, Sanjoy, Robi, and Shekhar—the four young men who take the train from Calcutta to Dhalbhumgar in Gangopadhyay's *Days and Nights in the Forest*—imagine the wilderness in a fairly conventional manner. They think that the forests will provide them with a clear break from life in the city characterized by strict ethos of civilization and culture. Their attitude is emblematized in Sanjoy throwing the newspaper away as soon as they reach the railway station at Dhalbhumgar. The newspaper, in that it is a chronicle of important incidents and events from the city and its surrounding areas, would always remind them during the sojourn that the break from the rules and regulations of the city that they may enjoy is only temporary. Hence, Sanjoy explains his action to his friends by claiming, "Damn it! I don't want to have anything to do with newspapers anymore" (Gangopadhyay 2010, 4).

Additionally, in their belief that the environment at Dhalbhumgar doesn't warrant following any rules and codes of culture or behavior, the men indulge in supposedly ritualistic stripping and unclothing during the second night of their stay in the forest guest house. Not only do they gain entry in the guest house on flimsy and ultimately illegal grounds in the first place, they imagine that giving up on their clothes will also reveal them in their "true and natural" states while in the forest. Except Sanjoy who is annoyed with Robi and doesn't participate in the exercise, the other three seem to behave as if stripping is the most natural act to be performed in the circumstance.

> In the twilight, Shekhar noticed that Robi was sitting in the nude, his clothes in a heap on the arm of the chair. He gave a loud laugh . . . . Ashim was rather amused. "He's right. Why bother with superfluities in the jungle at night . . . come let us all get undressed," he said, taking off his clothes and pulling enthusiastically at Shekhar's.
> "Stop! I'll take off my clothes myself." (Gangopadhyay 2010, 76)

The rather desperate attempts by the four men to achieve a break from the city as they travel into the forests actually provides the first point of criticism of their attitude in the novel. Gangopadhyay's narrator presents two scenarios where the city comes back to interrupt their journey. In fact, the first return of the city is apparently unconscious while the men themselves call on the city, as it were, in the second instance.

After having spent the first night in the forest guest house after bouts of drinking and merrymaking among the poor tribals in the modest marketplace of Dhalbhumgar, the four of them wake up to multiple requirements. Ashim is anxious to get Aspirin to check and get rid of his hangover from the previous night. As none of them is carrying a comb, Sanjoy and Ashim are entirely clueless about what they could do to bring their unruly hair into shape. And, even if they couldn't care less about reading newspapers in the forest, it just wouldn't do if they did not know the latest update about the test match between India and West Indies being played at Madras (now Chennai). They also need eggs and butter for breakfast and feel the urgent necessity to stock up on cigarettes as well.

Clearly, in an ironical and tongue-in-cheek manner, the narrator highlights the selective nature of the break from the city sought by the young men. They do not want to give up on those markers of city-life that makes existence comfortable and convenient for them but imagine the forest as a site where they can avoid all restrictive rules and regulations. They imagine the wilderness, in other words, to be thoroughly amenable to their own desire of control and appropriation. Unsurprisingly, as they witness the sunset in the forest while returning from the marketplace at Dhalbhumgar, they judge it to be the same or, at least in no way inferior, to those sunsets that are often seen in Westerns. They also regret that they don't have a camera to capture it.

> Crossing many narrow pathways, they entered a wide road in a clearing, only to find the sun setting in all its splendor. A rush of red breaking through the fleecy clouds had spread all over the sky, turning the tips of trees bronze in the play of light at the day's end. They had never seen such glorious sunset. They had seen it in films. They began to think of all the "Westerns" they had seen. "Do you remember *Garden of Allah*, with Burt Lancaster?" Robi asked.
> . . . "Everything looks attractive on film. The sunset, if captured on camera, could easily rival a Hollywood scene. Wish I had brought one along . . . . Shekhar just wouldn't let me."
> "Of course I wouldn't. We can't go about with expensive equipment in a place like this. We'd always be afraid of losing it. Since we don't have it, we don't have to worry about it." (Gangopadhyay 2010, 19–20)

It is in Sanjoy's annoyance with Robi as mentioned earlier that the second point of critique of human-environment relationship in *Days and Nights in the Forest* comes about. Sanjoy is angry at Robi because the latter has thrashed Lakka, the help that they had acquired on the way to the guest house. Lakka, a Santal young man, had aided the four friends by carrying their luggage and also getting them essential groceries, cigarettes, and country liquor. Robi accuses Lakka of fudging the accounts while buying such items and beats him black and blue. Apart from Sanjoy who is concerned about their safety in the forest and about a paternalistic sense of decorum and courtesy, the others do not find anything wrong in what Robi does to Lakka. And, the incident is symptomatic of their attitude toward the general population of Dhalbhumgar. The approach gets reflected in Shekhar's suggestion to his three friends that social revolution won't be possible unless the populace accepts city-bred, young men like them as its leaders.

> Shekhar said, "Do you realize we are aliens here? We are so different from these people in the way we walk, talk and behave. Who would say we belong to the same country? How can there be a social revolution if they don't think of us as one of them, if they don't do as we say? (Gangopadhyay 2010, 98)

In this instance, Gangopadhyay's novel is undertaking a vital critique of the ethics and politics underlying postcolonial India's nascent political formations. The suggestion seems to be working in at least two ways. First, that postcolonial India's polity hasn't been able to provide the bare minimum to its populations, especially to the most vulnerable among them. Second, that the country's independence has merely meant the replacement of India's colonial elites by the upper classes. And, the marginalized and subjugated continued to be ruled and suppressed, albeit by rulers with brown skins.

The critique of India's postcolonial polity vis-à-vis the sheer irresponsibility that it exhibits toward the country's marginalized is widened in scope and rendered in even sharper terms in Gangopadhyay's novel. The four young men engage the watchman of the guest house, Ratilal, in housework like drawing cold water from the well, cooking food and making tea, and cleaning and dusting the rooms, and so forth, even as he repeatedly reminds them that his wife is seriously ill and might not survive for long. Besides, they show no concern about the fact that he might lose his job if the Conservator of forests in that area discovers that they had been permitted to stay in the guest house without proper procedure being followed. Duli, the Santal tribal woman who asks the four men if she could do some work for them in the guest house, suffers even greater neglect at the latter's hands. Her case is symptomatic of the young men's understanding of Santal women as being readily "available" for (sexual) (ab)use. Shekhar and Robi's suggestion that they take one woman each when they see a bevy (Duli among them) for the first-time attests to this fact.

> Robi simply hadn't been able to take his eyes off the girl [Duli] in the blue-bordered sari.
> "Looking at them reminds me of the days when female slaves were sold in markets. These girls too seem to be waiting to be sold, waiting to be picked up by someone, sometime," Shekhar mused.
> "Come, let each of us secure one," Robi said (Gangopadhyay 2010, 34).

Robi is unable to check his desire for Duli and spots her after a trip to the village market with his friends. She tells him that she had come to the guest house looking for work as she hasn't had anything to eat for five days. She asks him for money to buy some food and, if there would remain something after satiating her hunger, then she says that she would like hair net and a red blouse for herself. As Robi agrees and gives only fourteen rupees to her, she feels a kind of benediction and is grateful for a young *babu* from the city comes to have some proximity with her. But, for him, the meeting is not only a transaction but it must also provide him with something. As he coaxes and cajoles Duli into making out with him, he believes that he experiences the fulfillment of his desire for "natural and pure organic pleasure" (Gangopadhyay 2010, 136–138).

It is important to remark at this point that Gangopadhyay's depiction of Robi's lovemaking with Duli in which the former conceives of her as a part of nature itself—organic, rustic, and so unlike the young women of an urban center like Calcutta—should not be construed as yielding to stereotypical representation. Gangopadhyay, in fact, tries his best in *Days and Nights in the Forest* to let the woman have control over her own sexuality. In this case, for instance, the narrative makes it clear that the sheer depravity of her circumstance leaves Duli with no option other than acceding to Robi's wish.[1] And, if the woman yields to the domination and control of the "outsider" patriarchal force, the novel would not even suggest that she is under control of patriarchal force that lies "within."

In what can be regarded as the postcolonial deprived answering back to the native elite, Lakka takes revenge on Robi by attacking him with the assistance of a few of his friends. In fact, the group injures Robi quite seriously even as Lakka justifies the attack as recompense for Robi having taken one of "their women." Ironically, it is in Robi's response that the reader of *Days and Nights in the Forest* registers the suggestion that Santal women like Duli don't belong to anyone and are not their "property." They have a will of their own and can chose to do whatever they like.

> "I'll do as I please," Robi shouted back. "I'll certainly beat up someone who is a thief. And what's this about 'our women'? They go to whoever they take a fancy to. Did I force her to come with me?" (Gangopadhyay 2010, 141)

If Lakka's fightback emerges as perhaps the most crucial way in which the four men touring the forests around Singhbhum get exposed to the limits of their understanding and control of the environment, Ray's cinematic adaptation of the novel uses a different set of strategies to make its viewer reach the same conclusion. The next section of the present chapter analyzes such means and techniques.

Ray's cinematic narrative begins with the four young men—called Ashim, Sanjoy, Hari (and not "Robi" as in Gangopadhyay's novel), and Shekhar—traveling to the Singhbhum forests in a car. They bribe the *chowkidar* Ratilal and gain illegal entry into the forest guest house. The act, however, doesn't dent their confidence at all as they continue to conceive of their sojourn as a necessary break from the forced civilizational and cultural values associated with the city-life of Calcutta. One of the first things they do after taking over appropriate spaces in the guest house is in fact to reach a collective decision that they would not shave their facial hair through the length of their stay.

The agreement implies that they think of the wilderness as a natural space far removed from civilizational ethos that would deem it polite and necessary to shave. And, as if to attest to the implication, Shekhar asks Hari to shout like Tarzan as they take a walk among the trees at night. Unsurprisingly, Hari obliges his friends by producing the characteristic Tarzan call, which goes something like "Mmmmm-ann-gann-nii" and continues to rise in pitch with each intonation (Dadabhai, *YouTube* 2014).

The film's narrative, however, sees through the attempts of the four young men to assimilate themselves in the milieu of the wilderness. It exposes them as being hypocritical in that they seek to retain their superiority as the civilized, urban-bred elite who will never be the equal of their tribal counterparts inhabiting Singhbhum and its surrounding areas. Shekhar, who Ray makes a fool in contrast with Gangopadhyay's Shekhar who appears to be calm and restrained, shouts at the children who turn up to look at him while he bathes by the well and refers to them as *asabhya*, literally, uncivilized. Additionally, when Duli comes to the guest house with a couple of friends looking for some work, Shekhar hires them on minimal wages and terms his action "tribal welfare." Through that apparently humorous evocation, he presents himself as the agent of the Government of India whose official duty is to undertake the welfare of the country's tribal regions and their inhabitants.

Unlike the relatively marginal importance given to the Tripathi family—constituted by the patriarch Mr. Tripathi, his daughter Aparna, the widowed daughter-in-law Jaya, and a very young grandson—in Gangopadhyay's novel, Ray's film endows the group with much greater significance. Aparna, in fact, emerges as the moral fulcrum around which the four young men visiting Dhalbhumgar come to realize the limitations of their worldview and the sheer irresponsibility of having thought that they can control and regulate the environment around them. The Tripathis are an elite family that occupies the only comfortable living space—a bungalow—in Dhalbhumgar and the four men have the chance to make their acquaintance while taking a walk in the forests on one night. The Tripathis understand correctly that the men are facing difficulties while staying in the guest house; therefore, they invite the gang to their house for breakfast the next day.

The meeting over breakfast serves as the opportunity for Ray's film to redeem Lakka of any doubts over his sincerity and honesty. While playing badminton in the courtyard of the Tripathi household, Hari drops his wallet. When he returns to guest house from the bungalow, he searches for it and can't find it. Immediately, he pounces on Lakka and beats him black and blue while accusing him of having committed the theft. The wallet returns, of course, the next day as it had been found in the Tripathi's courtyard. What Ray does in his film, in other words, is that he establishes Lakka as honest and straightforward in a concrete manner before showing him as attacking Hari with a stick in the forest.

Much like the redemption that the narrative of Ray's *Araneyer Dinratri* provides to Lakka, it makes a very strong point regarding Sanjoy's reading of Jaya as well. As he comes to know that Jaya has lost her husband recently to an alleged act of suicide, he comes to enjoy quite some proximity with her. Yet, he doesn't quite think of her as someone who might desire (him) sexually. During the time that they spend together, Jaya asks him, apparently casually, if he likes Santali ornaments. Sanjoy responds by suggesting that they look good when worn by Santali women. Later, as they meet in the Tripathi's bungalow over a cup of coffee, Jaya shocks Sanjoy by appearing in front of him all dressed up—her resplendent attire having been accessorized with heavy and very obviously noticeable Santali ornaments.

As pointed out by Ashis Nandy in his detailed psychosexual analysis of Ray's cinema including, of course, the film that concerns the present chapter, Jaya's appearance is meant to be seductive and erotic. Sanjoy is supposed to understand that widowhood doesn't necessarily entail that the woman stops having sexual desire or expressing her (erotic) likes and dislikes. Widowhood, in other words, doesn't mean that the woman should give up on all her agency and spend the rest of her life pining for the partner that she has lost. Clearly, Sanjoy is unable to anticipate this statement from Jaya and even as she breaks down immediately, he understands that his reading of Jaya, as a widow and as a woman, has been misplaced throughout the course of their meeting and subsequent interaction (Nandy 1995, 244).

However, as indicated earlier in this chapter, it is Aparna and not Jaya who comes to be the leitmotif through which Ray's film critiques the worldview of the four young men and presents it as hypocritical and insensitive. And, with regard to Aparna, it is Ashim who is shown in such critical light. While attempting to size her up and to establish her as just another "party-going type" girl from Calcutta, he asks her to show him the space within the family bungalow where she spends the most of her time. On hearing this, Aparna takes Ashim to a single room built at quite some height outside the bungalow. Once inside the room, Ashim notices that Aparna enjoys her solitude as she has great taste in books and music. She, for instance, reads Arthur Miller, Agatha Christie, and John Donne and listens to Vilayat Khan, Bismillah Khan, Mozart, and Andres Segovia. She, in other words, just doesn't seem to be someone who would spend a lot of time in following high-society mores and protocols of polite interaction.

The visit to the Tripathi household isn't the only time, however, that Aparna's sensitive behavior and approach cause Ashim to be ashamed of himself. Unbeknownst to him, Aparna has seen him bathing by the well without any clothes till his waist from the top. Moreover, Aparna and Jaya come to know that the Conservator of forests, a friend of Mr. Tripathi, had given the four young men two hours' time to go to some other guest house as they had no permission to stay in this one. The sisters-in-law have also seen them in a drunken state while they were returning from the marketplace at Dhalbhumgar. In their inebriated condition, they had caused the lengthy stoppage of Mr. Tripathi's car by blocking the road.

When the sisters-in-law visit the forest guest house to meet the four friends, they all play a memory game in which they are supposed to remember the sequence of names of famous personalities from history one by one.[2] Ultimately, as everyone else is out of the game and only Aparna and Ashim are left, the former deliberately loses by claiming that she has forgotten the sequence. Ashim understands this and brings it up while proposing his love to her as they meet outside the village market. On this occasion, Aparna informs him that unlike him, she has witnessed and experienced terrible sadness in her life. She recounts that she lost her mother to a fire accident and has been afraid of forest-fires ever since. Besides, her brother (Jaya's husband) committed suicide while in England. And, she suggests that Ashim, who hasn't lost anything or anyone to such accidents, will not be able to understand the pain felt when people try to put someone down even when they are in great distress.

Aparna's status as the moral agent in Ray's film gets completely concretized when she enquires of Ashim if the watchman was indeed going to lose his job because of the foolishness and brashness of the four friends. Moreover, she forces him to cross to the "other side" and see the pathetic conditions in which the watchman, his ailing wife, and children survive. Their habitat is an extremely small room in which the children sit surrounding their sick mother who tries to keep her high fever in check by moving a flimsy fan over herself. However, much to Ashim's relief, Aparna assures him that the watchman won't lose his job as she would discuss the case with her father who, in turn, will mention the same to the Conservator of forests.

All these instances reveal that much like Hari who misreads Lakka and Sanjoy who is absolutely incapable of registering Jaya's presence as a (sexual) being, Aparna's sophisticated personality also remains inscrutable for Ashim. In fact, the viewers of Ray's film begin to wonder if Ashim doesn't declare his love to Aparna at least partly for this reason, that is, his inability to pin down, control, and ultimately determine her personality. Besides, Aparna exhorts him to let go of his worldview that seeks to determine the environment and give into the very unpredictability and contingency that mark accidents. She, in other words, seems to be telling Ashim that accidents will thwart the best of his attempts to regulate the flow of things around him and that he must learn to take things as they come his way. Otherwise, there would be no scope for his personality to develop in a healthy and productive manner.

Aparna's advice to Ashim appears symptomatic of the critique that Gangopadhyay and Ray undertake in their respective narratives. Both present the four young men with scenarios that develop in ways that they could not have foreseen or predicted. And, the four of them are shown to be horribly misplaced wherever they try to make sense of what they witness and experience in the wilderness. Crucially, the environment—the forests with their inhabitants—not only becomes the site where their hypocrisy gets exposed but also comes to serves an ethical reminder to postcolonial Indian elite. The reminder is actually an ironical and sarcastic chiding that they have been unable to do even the bare minimum required by the country's most vulnerable populations.

What lesson then, as it were, do the two narratives hold for today's world? What do they tell us about the relationship between postcolonialism and the environment? It seems that both the texts demand an ethics of responsibility and care to develop between postcolonial subjects and the environment. Postcolonial citizens of today's rapidly globalizing world should not conceive of the forest, the wilderness, and their peoples as "objects" for usage and consumption. Instead, they need to remember that the latter represent the limits not only of our greed but also of our potential to bring about a change in the world. Consequently, we must dwell on the earth in a way that would always entail the recognition that we share something vital with the natural world—the forests, plants, and animals—and how we can get closer to imagining what is it that we share.

If postcolonial nations and their peoples do not pay heed to the environment, they would merely be replicating the practice of subjugating the (natural) other that the different colonial powers also undertook at different points of time. And, in such a case, the acquisition of independence and arriving at the postcolonial moment, as it were, would be rendered completely valueless—for what is the point of something as momentous as throwing away of the imperial and colonial yolk if we must continue with the same values that underlay the enterprise?

NOTES

1. It is very interesting to recollect at this point the recent controversy regarding Hansda Sowvendra Shekhar's collection of short stories *The Adivasi Will Not Dance* (2017). The book was banned by the government of Jharkhand for it was felt that it insulted the dignity of the Santali women of the state by implying that they indulged in sexual encounters with strangers because they *liked* to. In my review of

the book, I had indicated that much like Duli, the Santal woman Talamai in Shekhar's story "November is the Month of Migrations" sleeps with a *jawan* of the Railway Protection Force only because she is hungry and poor and finds no other way to sustain her bare existence. For details regarding the controversy as well as Shekhar's depiction of Santali life, see Thakur (2017).

2. For a detailed and insightful psychoanalytical reading of the memory game in Ray's *Araneyer Dinratri*, see Bandyopadhyay (2012). Bandyopadhyay's reading is important for the clear connections he is able to establish between the choices that different characters of the film make outside of the game and the names that they mention in the sequence. Needless to state perhaps but the choices that the characters make reveal a considerable amount of information about their sociopolitical and psychosexual makeup.

BIBLIOGRAPHY

Bandyopadhyay, Sibaji. 2012. "Ray's Memory Game." In *Sibaji Bandyopadhyay Reader: An Anthology of Essays*, 145–213. Delhi and Kolkata: Worldview Publications.

Dadabhai. 2014. "Aranyer Din Ratri (Days and Nights in the Forest)." *YouTube*, directed by Satyajit Ray. https://www.youtube.com/watch?v=Uu--1QfYKHc.

"Environment." n.d. *Lexico*. Oxford: Oxford University Press. https://www.lexico.com/definition/environment. Last accessed on 14th August 2020.

Gangopadhyay, Sunil. 2010. *Days and Nights in the Forest*. Translated by Rani Ray. New Delhi: Penguin Books.

Harrison, Robert Pogue. 1993. *Forests: The Shadow of Civilization*. Chicago and London: The University of Chicago Press.

Nandy, Ashis. 1995. "Satyajit Ray's Secret Guide to Exquisite Murders: Creativity, Social Criticism, and the Partitioning of the Self." In *The Savage Freud and Other Essays on Possible and Retrievable Selves*, 237–266. New Delhi: Oxford University Press.

Singh, Upinder. 2017. *Political Violence in Ancient India*. Massachusetts and London: Harvard University Press.

Thakur, Chinmaya Lal. 2017. "Stories We Don't Like to Hear: Review of *the Adivasi Will Not Dance*." *LittCrit: An Indian Response to Literature* 84 43 (2): 171–173.

Chapter 9

Postcolonial Ecology and Representation

Exploring "Ashani Sanket" as an Ecofilm

Neepa Sarkar

Today, more than ever, we realize the catastrophic consequences of human actions upon nature and the impending natural calamities that follow. And as the world battles COVID-19 pandemic, we realize that we are living on a planet where ecological disasters apart from socioeconomic inequality and ethnic clashes have completely engulfed us. Consequently, it becomes imperative to interrogate the present world order, which taking its cues from modernity has, on one hand, broadened the spectrum of the human mind and, on the other hand, brought in Industrial Revolution and European imperialism as Linda Hutcheon states in her 1974 essay, "Beautiful Losers: All the Polarities" (Hutcheon 1974, 43). If one seeks to imagine a post-imperialist, ecologically viable notion of society, then one has to question and reconfigure the place of humans in nature; for postcolonial studies cannot ignore environmental issues that were central to conquests and domination in its quest for justice and rewriting history. As one delves deeper into addressing the hierarchical binaries of power structures and racial exploitation, one realizes that the issues of animal and environmental studies also have similar concerns.

Using Gayatri Spivak's notion of "othering," Stephen Slemon has defined the discourse of colonialism as a structure that produces and naturalizes the hierarchical power systems of imperialism, making them appear inevitable. Imperialism employed both physical and mental tools to "domesticate" the natives and their land and transform them into "knowable" specimens to be categorized and tamed. This outlook, often spoken of in postcolonial studies, can be seen in early travel writings and literature too. Texts on travel writing and such accounts of circumnavigation and exploration complemented the projects of colonial expansion. So Shakespeare's Caliban from *The Tempest* is just a native who initially does not realize the dangers of imperialism that come to his island in the form of Prospero, a rereading of the play might suggest. Shakespeare's characterization of Caliban was based upon the popular travel accounts of his day in which distant lands and people were often portrayed as exotic, dangerous, foolish, and nonhumans. This unquenchable tendency to measure the whole world, to circumscribe, and chart the whole globe to a monocular perception came into its own during the Renaissance wherein cartographies were generated, sealing the link between geography, identity, and empire. Cartography and the politics of naming, mapping, and essentially categorizing the "Other" was promoted through colonization and disseminated through the voyage accounts of Christopher Columbus, Marcopolo, Vasco da Gama, James Cook, and others.

English literature of the sixteenth and seventeenth centuries is replete with such indigenous characters who need to be "known," named and disciplined, for instance, Man Friday in Daniel Defoe's *Robinson Crusoe*. With the rise of the Empire, the colonizers realized the need to build intellectual systems and power discourses, which would sustain and consolidate the imperial enterprise. Thus, "Ideological State Apparatuses" (Althusser) came into play, crippling the minds of the natives and developing a sense of inadequacy in them in the hierarchical binary relationship between the colonizers and the colonized; it has been extensively explored by Frantz Fanon in his *The Wretched of the Earth* (1963). Colonization also sought to bring unknown (read unmapped) territories and their flora and fauna into the Western knowledge structure and utilize it for their own progress and well-being. Postcolonialism and ecocriticism are inextricably interlinked and offer significant insights on how exploitation of nature is connected with the histories and practices of colonization. There are umpteen ways in which colonial history has displaced and exiled people from their land that to reimagine the notion of home and belonging becomes deeply fraught for the oppressed; both psychologically and in terms of landscape and nature which has been destroyed literally as well as imaginatively too. However, imagination seems to be the only way to recover land lost to colonization as Edward Said opines, "because of the presence of the colonizing outsider, the land is recoverable at first only through the imagination" (Said 1993, 77).

The European empire was constructed on a mercantile premise, which soon cascaded into a colonial assertion over land, people, and resources. The imperial approach consisted of a racial and gendered hierarchy of subjects (animate and inanimate) in the manner of the binary between nature and culture that often relegated the indigenous, non-Europeans and women to an object akin to "innocent" nature; and the white, landed, heterosexual male to rationality and culture. The figure of the white male explorer and adventurer often got extolled in the Western travelogues who would discover "virgin" lands of immense wealth to be conquered and utilized for the benefit of the empire. Such travelogues were replete with a deliberate reduction of the native knowledge systems and devaluation of indigenous modes of thinking that led most certainly to an obliteration of the ancient existence of the rich native culture and literature. Consequently, a postcolonial way of thinking regarding ecocritical issues would involve analysis of how empire has over the years constructed the human and the nature and their relationship and how it has semantically affected identity derivation and formation in the erstwhile colonies. Dipesh Chakrabarty in his *The Crises of Civilization: Exploring Global and Planetary Histories* (2018) speaks about issues of environmental justice, which cannot dispense of with the power relations and exploitation often associated with postcolonial studies particularly at a time when we all are jostling with effects of anthropogenic changes affecting the climate.

Postcolonial ecocriticism is an interdisciplinary field that investigates the discourse of colonization where often anthropocentrism and Eurocentrism are linked together, in a bid to justify the supremacy of the European imperialist and his colonial policies, which viewed the indigenous people and their cultures as primitive and animal like. Alfred Crosby in *Ecological Imperialism* (1986) speaks of the ways in which exchange of goods and ideas took place between the Old World and the New always in an unequal manner. A postcolonial approach to ecological issues is beneficial when connected with cultural and historical analyses along with ways that highlight the varied forms of epistemes and the conflicts between them. This approach cannot but look into power relations that affect nature at multiple levels from domestic to the global. Scholars and writers are increasingly becoming aware regarding the ecological vulnerability, which was a result of experiences specific to colonization and exploitation of indigenous resources. Peder Anker noted in an essay (2001) that the beginning of ecological awareness corresponded with the end of British imperialism. For the colonizer, indigenous lands become represented in the semiotic order as primitive and dangerous and hence needs to be domesticated and enclosed in a neat categorical order, available for social and economic control to the colonizer. The native land in the beginning is often like the "Lacanian Real" for the colonizers, difficult to understand and measure and yet whose presence can never be completely ignored. In fact, it hovers and haunts the colonizer's notion of self and identity engulfing him in an anxious state of mind and he doesn't rest or find relief till he has tamed and included it in the colonial scheme of exploitation, obliteration, and extraction of resources.

Crosby's term "ecological imperialism," is seen as a broad term encompassing brutal colonization of native lands to introduction of nonindigenous animals and agrarian practices to the colonized lands; it may also be associated with the idea of environmental racism as defined by the American philosopher Deane Curtin as a link, in both theory and practice, between the oppression of one that has an impact on the other; an environmental based discrimination meted upon the marginalized (both economic and social). In India, this environmental discrimination was often linked to caste hierarchies, deciding in a structural way, which community gets access and how much do they get access to natural resources, like water, as often portrayed in Munshi Premchand's (1880–1936) short stories.

Val Plumwood, an ecofeminist, sees environmental racism as the ultimate form of "hegemonic centrism"—a self-privileging viewpoint that is common to all the discriminatory—isms and has historically exploited nature, claiming earth as the sole refuge of only the humans. Plumwood further states that this centristic notion is built on the binary idea of the human and the nonhuman, which finds ramifications in the Western rationalization of imperialism, viewing anything (animate and inanimate) non-Western as vacant and void and the indigene as primitive, childlike and close to nature; quite oxymoronic as the land, which the "ignorant" and "primitive" indigene inhabits, is often seen as exotic and dangerous, bordering on the desires of colonialism and unless brought into the imperial fold, rejected as literally an empty space to be filled in. Val Plumwood further opines that the structural processes of our cognition and comprehension of the world is often based on the binary power dynamics of categorizations and Othering and that it extends to the way we view and treat environment too. And that ecocritical attitudes and studies must address this injustice, which also sees nature as continuously providing for mankind. For if we continue in this path of destruction, we are simply replicating the colonizer's discriminatory ideologies on a planetary level. A postcolonial literary approach toward environment should engage with visual representations of nature and examine the role of nature settings in literary texts. Literary ecocriticism, further, emphasizes upon the link between the social and the eco sphere, civilization, and nature and how human actions and decisions affect both. Writers have often made us aware of the increasingly dismal and alarming planetary conditions that we continue to live in if we keep on this path of utilitarian approach toward nature and the need to redress it through books like Charles Dicken's *Hard Times* (1854) or Rachel Carson's *Silent Spring* (1962). The present epoch has often been described and discussed by Crutzen and Stoermer as the "Anthropocene" and thus a need to critically engage with the approaches that locate and entrench our sociocultural and ontological experiences with those of the environment. These will include different visual and narrative methods that may represent the discourse of postcolonial ecology.

Since we live in an ocular-centric world where we are every moment bombarded with a sea of images that may or may not leave an imprint on our fragmented minds, epistemologies of climate variations and global ecology and ideas regarding planetarity often get represented via the visual medium of films. Cinematic ecocriticism incorporates ecological considerations into our viewing and perception of cinema. Ecofilms or ecocinema expressively associates with concerns regarding nature by focusing on issues of ecological justice or by making landscapes and fauna the prime reference point in the narration. Ecofilms also may present films in a philosophical manner forcing us to interrogate our positions and actions as inhabitants of this planet. Ecocinema often focuses on issues of understanding the ecosystem and eliciting a nonhierarchical response and relationship between nature and man. In postcolonial countries, one must always take into account how the violent histories of colonization have molded our ways of perceiving and comprehending the environment and our relationship with it.

Ecocinema remains a broad term inclusive of all sorts of environmentally conscious representation either emphasizing our need to be aware of the present concerns that affect our environment or to understand the need to address the wrongs of the past committed toward environment by the colonizers. This chapter closely explores Satyajit Ray (1921–1992) directed "Ashani Sanket" or "The Distant Thunder" (1973) based on the novel by Bibhuti Bhushan Bandopadhyay and tries to bring to attention the specific ways in which this visual text has communicated to the audience on issues of famine and colonialism. Visual culture, as we know, aids and maintains specific global discourses and perceptions. Cinematic ecocriticism urges us to include ecological reflections of the times and accept our role as producers and consumers of films who inhabit with both animate and nonhuman forms upon this planet in an intricate symbiotic relationship in the process of evolution. One of the earliest texts that provided an ecocentric analysis of cultural texts was Jhan Hochman's *Green Cultural Studies: Nature in Film, Novel and Theory* (1998). Hochman asserted in his writings that one has to include nature in cultural studies for culture is ingrained and reliant on nature.

The way we view or analyze a film with an ecocritical lens has its precedent in literary ecocriticism whose early proponents Cheryll Glotfelty and Harold Fromm defined it as an earth-centered and environmentally aware method to analyze texts and explore the connection between texts and their material settings. Cinema is just not spectacle and storytelling but becomes a reference point from which meaning can be derived. It is a sort of "poiesis" (Heidegger) and brings forth a world, which can concur with reality or prompt the world to action. Satyajit Ray, it seems to the researcher, often appears as an eco-auteur in many of his films, whether it be the child Apu in "Pather Panchali" glancing at a passing train through the Kash flowers-filled field or the opening scenes of "Ashani Sanket" where Ray chooses to bring to life a village through the sounds and views of its natural surroundings—lush green fields, chirping birds, and brimming ponds. The film presents the village as an idyllic, pastoral place where nature and man seem to be cohabiting in harmony and balance. However, the tempo soon changes as is reflected both visually, in the form of dark clouds and aurally with the looming sound of distant thunder as well. The story is set in the backdrop of World War II and the 1943 Bengal famine in which almost three million people died of starvation. The hovering presence of World War II in the movie is portrayed by fighter jets flying over the village but the people seem to be unperturbed and undisturbed by them and in fact rush to see them in the sky, which points to the security that the village seems to be enjoying for the time being. War seems to be a distant idea in this remote idyllic village and people go about with their daily chores in a routine manner bereft of anxiety. References to epidemics (cholera) and water scarcity also abound from the beginning in the film when the protagonist's wife talks of her stay in a previous village, which faced water crisis. She shares an affable bond with the other women of the village and while swimming in the local pond she speaks of her stay in a previous village where water was scarce and getting to swim in it was a luxury. With a playful banter among the characters Ray successfully places a sense of anticipation in his audience who if keenly observant understands that the first signs of trouble will soon disrupt the idyllic atmosphere of this village.

The village setting is embedded in caste hierarchy where access to resources is based on one's status in society and the protagonist Gangacharan is shown as the only Brahmin in the village who has taken up the role of priest, teacher, and doctor all rolled into one. His caste status makes him highly respected and valued in the village and no one questions his competency and knowledge to do the things as he says he can. He is often offered good food, in the form of grains or fish, for his priestly services. Being a shrewd man, he quickly decides to increase his income by expanding his role in the village and with the help of the important people of the place he decides to open a school, telling the illiterate villagers that they must send their children to school so that they can have a better future. In a scene, Ray shows his protagonist trying to show off his knowledge of Sanskrit but falls short, which only the audience is able to gauge and not the other characters in the movie. In another scene, reference to World War is deftly brought in when mention of rise in the price of rice is made and it is also discussed that since it is coming from Singapore, which is captured by Japanese, availability is less. When the protagonist, seen as an erudite person, is asked where Singapore is, he replies it must be close to Midnapore (a district in Bengal). Again, the audience is let privy to the faults and pretensions of the protagonist. However, he is not shown as a bad or an evil character but as simply a young man who has his faults and who knows how to make use of his caste in order to procure material things so that he and his wife can have a comfortable living. His wife

Ananga is an amiable lady and mixes with the women of the village though within the limits of caste. She is referred to by her caste "Baman Didi" or Brahmin lady and nobody calls her by her name as such.

The somber anticipation of what is to come appears in the form of an old, hungry priest of a neighboring village who calls out to the Gangacharan after he sees him coming back from another village loaded with food and other things, which the villagers had given him after he performed his priestly duties there. The village was under an attack of cholera, and the villagers pool in money to perform a ritual so that they can be cured of this malaise. Gangacharan performs his duties to the utmost and while leaving advises the villagers on matters of health and hygiene, particularly asking them not to drink water from the local pond for some time. This shows the intrinsic amalgamation of old traditional methods and new ones in the form of incorporation of science into daily lives in the character of Gangacharan. Initially, the Brahmin priest Gangacharan and his wife are able to live comfortably as work and money keeps flowing in for him as he goes about utilizing his caste privileges but soon due to less import of rice, its availability decreases and price increases as everyone is not able to procure their required need of it. Consequently, there is loot in the granary of the rich rice farmer of the village, where the protagonist finds himself pushed and forcefully shoved aside, which devastates the priest as he feels humiliated and disrespected despite his caste status. This is the first time he has faced such humiliation and he realizes that in front of want and hunger, caste privileges and benefits will not hold. This makes him and the audience understand the basic wants and needs of man is primeval and few. Scenes of famine and impoverishment soon starts to percolate on to the screen and in an instance the protagonist's wife, Ananga finds herself staring in a shocked manner upon the people of a neighboring village who are in search of food, standing in front of her house and asking her for some starch. As days go by, the price of rice keeps on increasing till one day the newspapers report that rice has disappeared from the markets. Appalled the villagers surround the grocer man of the village who used to have a granary stocked with rice grains. However, the man tells the villagers that he has no grains in his granary and gives them the key to go and check. The villagers finding nothing leave his premises in a skeptical mood, accusing him of hiding the rice in his house. Later, he is attacked by a group of men and rendered unconscious and while visiting him in the capacity of a homeopathy practitioner, Gangacharan overhears him asking his daughter if the villagers had looted that small amount of rice that he had hidden in his house.

The film constantly manifests ecological sensibility and equates women with nature, showcasing the resilience of both. In the film, one will often find the character of Chutki depending on nature as she goes in search of food when it no longer remains easily available. She brings news about food to the priest's wife Ananga and takes her to the forest in search of wild potatoes. Also, she is forced to sell her body in exchange of food to the local disfigured worker of a brick kiln who skillfully seduces her over a few grains of rice. She later tells Ananga that in front of hunger she forgets reason and all morality. Hunger and body get intimately connected, reducing humans to an almost primitive status, forgetting their dignity and integrity for a few morsels of rice. Ultimately, Chutki is shown as leaving the village with the disfigured man for the city in search of food. This becomes a subtle commentary on the notion of urbanity and how colonization wreaked havoc with brutal and mismanaged policies affecting the lives and livelihood of the indigenous population. The film in unique ways is able to highlight how greed and desire brings in destruction and ultimately no one can save oneself when hunger and starvation takes everyone in its throes.

Gangacharan, secure in the hierarchical system of a caste-based society, understands the working of this small village society and in exchange for the services given by him as a priest and physician demands material rewards. In an instance, he goes to a neighboring village where a bout of cholera has erupted and the villagers want him to conduct a ritual to rid the place of cholera, which he performs along with giving them advice on how to maintain hygiene, which is a clever combination of science and religion through the character of Gangacharan. Soon the tranquillity of Gangacharan's village is destroyed as the ramifications of the War get felt even in the remote areas of the colonies as nature is thrown awry and an imbalance sets in. The famine tears apart the civil fabric of the village society, turning the men into monsters, shredding their solidarity and brotherhood, whereas on the other hand, women come together strengthening their camaraderie as they rummage for food in the forest. Where civilizational methods fail, humans turn to nature to procure food is highlighted in the film. The film moves from scenes of plenty in the beginning to scenes of scarcity as the film ends depicting the effect of the 1943 Bengal Famine and which is seen as manmade and artificial and in which many perished. Gangacharan gradually starts understanding and questioning the sociocultural system that he is a part of and toward the end tells his wife that if they have to do odd jobs despite their status, they will do it together. The film builds up on contrasting scenes of greenery, abundance, and serenity of nature and the greedy, consuming nature of humans and the tragedy of what is to come seems inevitable.

The food crisis that erupts soon engulfs the small village of Gangacharan and prices of food commodities soar up. The incorruptibility of the village becomes shattered and men resort to plundering and violence while the women of the village are forced to scourge for snails in the ponds. Nature stands tall and indispensable in the film, a character in itself affecting and molding the life of all the characters of the plot. The film in subtle ways is a strong indictment of war and casteism at the same time to a keen viewer reminds that human life is often dependent on environment. The novel as well as the movie made a realistic depiction of the famine that Bengal faced under British rule. The film alludes to famine, hunger, and starvation throughout and though the Bengal famine did not receive much academic attention, it however found representation in literature and other artistic works. The film analyzes human relations in times of scarcity and impending doom in the face of a manmade disaster, that of the Bengal famine as well as giving the audience a delineated caste study of a village society. Ananga remains incorruptible till the end even in the face of hunger, and in the last scene when she gets news of a person from her village lying unconscious some distance from her house, she rushes to help but is unable to touch her as she is an untouchable. However, she provides her with food but it turns out to be futile as the woman dies unable to eat it as she is too weak. The food is taken by another girl who had been eyeing it since Ananga gave it to the lady.

One striking instance that stands out in the film is when Gangacharan visits a nearby village in search of food, replicating the same scenario, which the older priest had been forced into. There the village grocer tells him he has rice only for his family and he would not be able to spare any for him. However, as Gangacharan is a Brahmin, he invites him to lunch at his home. Being a Brahmin, Gangacharan cannot have food cooked by persons of other caste so instead the grocer provides him with all the ingredients and Gangacharan cooks food for himself with the aid of the grocer's widowed sister. While cooking he remembers his wife who will be waiting for him to bring some food. Seeing his forlorn expression, the grocer's widowed sister promises to pack the food for his wife too and also give him some rice grains from her share. The film expresses a valuable generous and self-sacrificial bond between women and nature.

The film provides us with visual representations to help us comprehend the spatial and temporal arrangement that shape our thought processes and actions. In *The Three Ecologies*, Felix Guattari defines environmental, social, and mental ecologies. Guattari was influenced by anthropologist Gregory Bateson who by combining the effect of ecology, cybernetics, and general systems theory saw mind as an interactive structure, involved in exchange of information, and is not simply located within but remains immanently circulating in the discursive dynamics of the global order. Postcolonialism has been critical of European imperialism and the colonial enterprise and mostly concerned with political and ethical issues but has realized its connection with environmental issues in times of debates on planetarity and sustainability.

Literatures of the world have through the ages spoken about the ravages of industrial societies and man's incessant greed for unlimited and unmitigated development often comes at the price of destruction of nature. There is a need for new literature dedicated to the purpose of postcolonial ecocriticism, which can bring about action and discursive critique into the field of cultural studies. Colonial history has not only physically displaced people but has created a mental alienation in the people regarding their own land and history and therefore ecological recognition of the injustices done against nature and its recovery is precarious. Postcolonial environmental representations should engage, and they do so, with violent legacies of material and cultural transformations affecting the environment beyond repair.

The researcher has tried to analyze the film through the lens of critical ecologies, trying to engage in a critical reading upon discourses on race, gender, and the injustices prevalent in the structure of colonialism. Critical ecologies as a method remains concerned with notions of domination, alienation, and ethics, and tries to address the issue of ecological misuse that have been the inheritance from imperialistic exploitation. As such films are also part of the process of production and consumption in the "cultural circulation" model, which was developed by Stuart Hall and other Birmingham School theorists. This model posits that the "cultural circuit" flows through moments of production of cultural products and texts when meanings get embedded into them. Using the idea of Encoding/Decoding and blurring the strict categorization between producer and consumer in the discourse of cultural text production and reception, the Circuit of Culture looks at process of communication, which includes production, representation, consumption, regulation, and identity. These processes work together in a collective cultural space in which meanings get created, molded, and reshaped. The film utilizes "conceptual maps" (Stuart Hall) that erstwhile colonies inhabit themselves in, trying to understand the processes of belonging, identity, and a postcolonial culture.

Extending Du Gay's argument, that in late modern societies, the economic and the cultural are hybrid category and cannot be sorted into different segments, in the film, one realizes that colonies facing the end of the rule of the empire had to undergo a lot of struggle to survive and to subsist. The Circuit of Culture is an important analytical model to understand the processes of culture and identity. This model can be used in the interdisciplinary field of postcolonial ecocriticism to understand the process of identity formation and derivation and explore the intimate connection between man and nature. Ecofilms or ecocinema as a subgenre in this field can aid in this process of understanding how nature too has been historically marginalized and wronged in the turbulent practice of colonialism. Satyajit Ray stands out as an eco-auteur who in many of his films made nature a strong but silent protagonist, broadening the imagination of man and at the same time facing annihilation in his hands. This film specifically addresses the issue of destruction of natural resources and utilizing the resources of colonies to cater to the imperial might and strength in times of World War, thereby leading to a manmade disaster in the form of famine and historically scarring the people (indigenes) and nature as well. The film in certain ways becomes a study of the calamity of famine and explores the concept of humanity and spirit in trying times.

BIBLIOGRAPHY

Althusser, Louis. 1995. *Ideology and Ideological State Apparatuses*. London: Verso.
Anker, Peder. 2001. *Imperial Ecology: Environmental Order in the British Empire, 1895–1945*. Massachusetts: Harvard University Press.
Chakrabarty, Dipesh. 2018. *The Crises of Civilization: Exploring Global and Planetary Histories*. New Delhi: OUP.
Crosby, Alfred. 1986. *Ecological Imperialism*. London: CUP.
Defoe, Daniel. 2018. *Robinson Crusoe*. New Delhi: Penguin.
Guattari, Felix. 2000. *The Three Ecologies*. London: Athlone Press.
Hall, Stuart. 2016. *Cultural Studies 1983: A Theoretical History*. Durham: Duke University Press.
Hutcheon, Linda. 1974. "Beautiful Losers*:* All the Polarities." *Canadian Literature* 59: 42–56.
Plumwood, Val. 2002. *Environmental Culture*. London: Psychology Press.
Said, Edward. 1993. *Culture and Imperialism*. London: Chatto and Windus.
Shakespeare, William. 2019. *The Tempest*. London: Wentworth Press.
Slemon, Stephen. 1987. "Monuments of Empire: Allegory/Counter-Discourse/Post-Colonial Writing." *Kunapipi* 9 (3): i–16.

Chapter 10

Land, Labor, and Family
The Impact of US Colonization on Puerto Rico in Esmeralda Santiago's When I Was Puerto Rican

Renée Latchman

Bursting with landscape and cultural references to her homeland, Esmeralda Santiago narrates to the world the beauty of her country, the richness of her heritage, and the flora and fauna of her island home, Puerto Rico, in her first memoir *When I Was Puerto Rican* (1993). As the title suggests, the story details Santiago's life before she left Puerto Rico. A significant contribution of Santiago's text is her presentation of the Puerto Rican environment of both the city and the country. Readers get a true sense of the ecological situation as she and her family move from place to place during the 1900s, leading up to their migration to the United States toward the end of the memoir. Importantly, Santiago historically contextualizes her narrative by including personal accounts of US invasion and colonization of her land and culture, with specific references to programs implemented by the US meant to erase, yet uplift Puerto Rican society in the name of aiding their country. Environmental theorists call the byzantine relationship between America and Puerto Rico—settler colonialism—which, according to Kyle Whyte (2018) in "Settler Colonialism, Ecology, and Environmental Justice," "refers to complex social processes in which at least one society [such as the US] seeks to move permanently onto the terrestrial, aquatic, and aerial places lived in by one or more other societies [such as Puerto Ricans] who already derive economic vitality, cultural flourishing, and political self-determination from the relationships they have established with the plants, animals, physical entities, and ecosystems of those places" (134–35). Still, the United States was not the first to impose settler colonization on Puerto Rico. Spain preceded the United States, coming in contact with the indigenous peoples—the Taínos—as early as 1493. Then, the Spaniards were more interested in amassing the gold of the land than colonizing it through occupation. Unsurprisingly, they exercised imperialism in the sixteenth century, making Puerto Rico a colony of Spain. Barbara Deutsch Lynch (1996) in "Caribbean Environmentalism: An Ambiguous Discourse" notes:

> Cuba, Puerto Rico, and Hispaniola were all characterized by extremely early settlement and decimation of indigenous populations, repopulation with Africans and Canary Islanders, persistence of a Spanish colonial presence well into the nineteenth century, a two-phased reconfiguration of the agricultural landscape to accommodate artisanal and later industrial sugar production, and direct U.S. military occupation to support the second phase of this reconfiguration. (230–31)

Therefore, the Puerto Rican landscape was already familiar with interference from other cultures who believed they had better use of the topography by way of government documents that cater to the colonizer's expectations, but which blurs the understanding of the colonized. In accord with Lynch (1996) regarding the challenge of acquiring authentic information about the environmental impact or injustices incurred by settler colonies, this research offers Santiago's memoir, *When I Was Puerto Rican*, as a valid source for gaining understanding about the "fundamental environmental questions and preferences" regarding Puerto Rico's colonization, which Lynch believes are "most precisely and accessibly expressed in fiction, poetry and narrative in art, in the built landscape, and in the theater of social movements" rather than in government documents (229–30). Therefore, via *When I Was Puerto Rican*, I will discuss the environmental injustices of US occupation/invasion through the lens of settler colonialism theory and demonstrate how settler colonization impacted Puerto Rican livelihood, food, education, and politics, pushing thousands to emigrate in search of better economic opportunities.

SETTLER COLONIALISM AND PUERTO RICO

Thousands of years before the European explorers arrived, there were indigenous peoples originating from the mainland of Central and South America who had settled in the Caribbean islands. These indigenous peoples developed distinct communities and political systems—a fact maintained by several Caribbean historians such as B. W. Higman, Franklin W. Knight, C. L. R. James, and George Lamming. By 6,000 BP, Higman (2011) claims, migration from Central America began to take place among the "archaic hunter-gatherers" who shared much in common with the hunter-gatherers from South America, and the Central American migrants moved swiftly across the coasts of Cuba, Hispaniola, and Puerto Rico (16). Later, these migrants, known as the Taìnos, Siboneys, Caribs, and Arawaks, became the indigenous peoples of the Caribbean before the arrival of the Europeans. The indigenous peoples met the European explorers in the fifteenth century when the Europeans landed on the island chain. However, many of these indigenous peoples died a few short years after European contact, succumbing to disease and maltreatment (Higman 2011, 76). In describing the colonial event, Higman (2011) declares, "European, colonization was . . . invasive. People already occupied the islands. . . . Only by pushing aside, removing, enslaving, or killing those people could European colonization make space to succeed" (52–53). Inadvertently, Higman, like many other post-colonial scholars in relating the colonial experience, was describing what is now known as settler colonialism, a structure, posits grounding settler colonialism theorist Patrick Wolfe, rather than an event. To further explicate settler colonialism, Tate A. LeFevre (2015) in "Settler Colonialism," adds that settler colonialism as an act is "premised on occupation and the elimination of the native population, while colonialism is primarily about conquest" (1). Evidently, settler colonialism is an ongoing activity that permeates all facets of the colonized's life; yet, settler colonialism implies that colonialism is a one-time event that does not necessarily include erasure of the contact people or their culture.

In the case of Puerto Rico, their original indigenous people were killed, but a new indigenous population was created by Spanish occupation and the insertion of African labor. It is this newly created indigenous peoples of Puerto Rico whose land and culture are later recolonized by the United States. According to Whyte (2018), once the "process of settler colonialism takes place or has already occurred in some region, the societies who are moving in or have already done so can be called 'settlers', and the societies already living there at the beginning of settlement, 'indigenous peoples'" (134–35). In this regard, the jibaro, peasants, and farmers of the Puerto Rican countryside, who characterize much of Santiago's memoir, are the new indigenous peoples, since the Taìnos, who are the original natives of Puerto Rico, were decimated.

A further distinguishing feature between colonialism and settler colonialism, claims LeFevre (2015), "is a distinct imperial formation. Both colonialism and settler colonialism are premised on exogenous domination, but only setter colonialism seeks to replace the original population of the colonized territory with a new society of settlers (usually from the colonial metropole)" (1). Thus, when the United States invades Puerto Rico, they aggressively begin to indoctrinate the population with American ideals as a method of replacing or transforming the population to mimic American society.

WHEN PUERTO RICO MEETS AMERICA

Shortly before US colonization, Puerto Rico gained autonomy from Spanish rule over its produce and could determine its own foreign relations with the hope of establishing its own government. Their desires were delayed, however, by US possession of the land, even though some Puerto Ricans were hopeful for the prospects of American rule based on the promises meted out by the American government. Becky Little (2017) quotes noted history professor, Lillian Guerrera, who states "when the Americans arrived, General [Nelson] Miles issued, very famously, a decree manifesto in which he promised to protect the life, liberty, and happiness of Puerto Ricans, and their property." Consequently, many of the poor, working-class Puerto Ricans who were peasants "took this as an invitation to side with the Americans in what was still a war against Spain" (Little 2017). Instead of expansion of local companies, however, many Puerto Rican landowners lost their land and "economic power to North American companies which came to own the principal centers of sugarcane production" (Marisabel Brás n.d.). Carmen Teresa Whalen (2008) in "Colonialism, Citizenship, and the Making of the Puerto Rican Diaspora: An Introduction" corroborates Little's findings, noting that after the United States colonized Puerto Rico in 1898, it was solely interested in developing the land for commercial purposes. Whyte (2018) calls America's manipulation of the land environmental injustice as they, according to Lynch (1996), transformed the "landscapes into endless oceans of cane [that] came to be associated with progress, order, and the whiteness of refined sugar, while root-crop cultivation on steep, inaccessible hillsides was associated either with an increasingly Arcadian pre-Columbian past or with Africanness-darkness, disorder, and rebellion" (231). Undoubtedly, America forfeited the promises they made to the Puerto Ricans, seeking only to endorse their colonial system and use Puerto Ricans as cheap labor.

Importantly, American settler colonization was significant in establishing a large working class based on their capitalist system. César J. Ayala (1996) in "The Decline of the Plantation Economy and the Puerto Rican Migration of the 1950s" expounds on the effects of US occupation on the Puerto Rican countryside and on the social class system. Ayala argues that the United States was actively proletarianizing the country by creating a monoculture crop without developing any other economic entities. Ayala (1996) further purports, "under the conditions imposed by the colonial relationship to the United States, rural proletarianization was accompanied by the establishment of sugar mills which in effect functioned as industrial outposts in the countryside, but no significant development of urban industry or industrial towns took place" (63). In other words, while profiting from the sugar industry, the United States did not develop any other areas of Puerto Rican commerce.

It is in this environment when the demand for sugar fell, Santiago's characters are located. Puerto Rico had no other form of business to support them, and thousands of workers were left unemployed creating a large working class and a drastic increase in unemployment. Moreover, the monoculture of sugarcane did not profit the country or modernize it. Rather, it created an imbalance in the economic structure that stunted development and industrialization, crippling Puerto Rico's economy. The effects of the stunted environment and economy are evident in Santiago's memoir through several of the characters who comprise the lower working class—having menial jobs or doing whatever they can to make money. Don Berto, Negi's neighbor, wields a machete but does not go out to work because there is no plantation to work on and manual labor is all he knows. Negi's father has no clear occupation, but he does physical labor that includes fixing houses; Negi's grandfather lives in the city but he does not have a formal job. He sells oranges to passers-by because his known occupation is on the land to which he no longer has access. Negi's mother also does not have a formal employment, and when she seeks one, what is available to her is doing laundry and cooking food for the upper class, since there are little to no jobs for women or uneducated lower-class citizens. Prior to US occupation, many of those persons who constitute the lower class could work and eat off the land even if they did not have formal jobs. However, under US rule, the land was no longer available to the people and the United States did not create any other forms of employment to facilitate the growing working class or to feed them and their families.

America's lack of commitment to the development of Puerto Rico and their land displaced many people including Negi and her family. As settler colonials, the US was acting in accordance with their imperial identity. Certainly, constant relocation became common among Puerto Ricans while under American rule. Intra-island migration became the norm based on job opportunities and the need to survive and provide. Negi and her family moved several times from the city to the country and back based on her parent's ability to provide for them and to find work. A major problem for Puerto Rico was that the United States did not value them as individuals and saw them only as labor. "U.S. colonial administrators viewed Puerto Ricans as incapable of self-government and as a pliable labor force" writes Whalen (2005, 4). This point of view emboldened the US in advancing Puerto Rico to become "a classic monoculture colony, producing one crop, for export, to [only the US] market" (Whalen 2005, 7). Obviously, this move was only beneficial to the US since subsequent to Puerto Rico becoming a monoculture colony, "land concentration and an economy based on crops" limited a family's ability to provide for their loved ones. Workers were unable to produce food for their homes and were forced to consume imported food from the United States

(Whalen 2005, 7). This was just the beginning of a number of processes meant to dominate and control Puerto Rican society.

Thus, to gain their livelihood, many people were forced to move from one locale to another. Forced relocation, according to Whyte (2015), is another integral tenet of settler colonialism (133) practiced by the United States. This type of relocation involves removing people from their original ecological space and creating a new space for them that does not necessarily cater to their ecological needs; or, simply removing the people altogether from their natural space without providing an alternative. Once the United States shifted its focus from sugar to other industrial products, the sugar mills became redundant and farmlands became empty plots of land yet inaccessible to Puerto Ricans for uplift or produce because the land belonged to the Americans. Thus, many of them were forced to leave the rural areas that housed the sugar mills in search of jobs or create new communities in the rural areas. Santiago provides a fitting example of an impending forced relocation in her text. Negi, the protagonist, and her family move to a barrio in Macùn where their zinc home shares a boundary with a finca or farm. Negi was certain the farm was owned by their neighbor, Lalao, a fellow Puerto Rican. However, Doña Lola, their jibara neighbor whose family was displaced by US occupation, corrects Negi, explaining that the farm is owned by an Americano, Rockefela, who is "going to build a hotel back there" (Santiago 2006, 55). In shock and concern for the land and the animals, Negi asks, "What will they do with all those cows?" Doña Lola responds, "You're worried about the cows? What about us? . . . Do you think they will let us stay here if they build a hotel?" Negi enquires, "Why not?" (Santiago 2006, 56). Doña Lola responds with a Puerto Rican aphorism that Negi understands as Doña Lola's expression of mistrust in Americans from past experiences. "I know the bull that charges and the serpent that stings" (Santiago 2006, 56). The unsettling of the indigenous Puerto Ricans through settler colonization always takes a toll on this population, so Doña Lola's expression that metaphorically ascribes the forcefulness of the "bull that charges" and the poisonous and hurtful "serpent that stings" to the act of colonization emphasizes the dangers to Puerto Ricans when colonizers repeatedly force them out of their environment.

While settler colonization boasts of creating jobs for the indigenous people of Puerto Rico, it creates white supremacy and low wage laborers even when the colonizers no longer need Puerto Ricans to farm their lands. Dean Itsuji Saranillio (2015) discusses an ideology suited to the actions of America in "Settler Colonialism" asserting that "primitive accumulation—the inaugural moments of violence that set the conditions for capitalism to exist—is a useful concept in examining the complex relations within settler colonialism structured by white supremacy" (291). This process of "accumulation by dispossession is an ongoing one as illustrated in Santiago's memoir, and it divorces a people from the means of production—from their ability to provide for the basic necessities of life—and they are then forced to live through wage labor" (Saranillio 2015, 291). Indeed, the large sugar estates established across the island produced a massive population of farmers, which eventually transformed into the large lower-class strata of Puerto Rican society that came to depend on wage labor once sugar lost its popularity—a move that severely scarred farmers and peasants like Negi's Abuelo (grandfather), and Negi's neighbor, Don Berto, grandfather to Negi's best friend, Juanita. Santiago (2006) describes Abuelo as "a quiet man who walked with his head down, as if he had lost something long ago and was still trying to find it" (94); and Don Berto with "hands . . . always moving, always searching for someplace to land" (49). Their presence in the memoir is significant in representing the dying community of people who constitute the farmers, jibaro, and peasants who were impacted by US capitalist ventures in the first few years of their invasion. The death of both men later in the memoir suggest the death of an integral group of people who were left to fend for themselves and who received no support from a government who promised to protect them and their property. Both men are symbolic of loss and a dying culture. Irrefutably, the United States never kept their word to "protect the life, liberty, and happiness of Puerto Ricans, and their property" (Little 2017). So, despite claims that Puerto Rico's economic crisis was a result of its overpopulation, Whalen (2005) strongly disagrees postulating that it was America's mismanagement of the land and its wealth that lead to Puerto Rico's economic demise.

In Santiago's memoir, an entire chapter is dedicated to the actions of the United States to infiltrate and transform several facets of Puerto Rican life. Whyte (2018) notes, the "settlers' aspirations are to transform Indigenous homelands into settler homelands" (135). To accomplish this, America must insert their values and beliefs in every area of Puerto Rican life. As the title of the chapter, "The American Invasion of Macùn" suggests, America invades Puerto Rico, but not through war, since this time, Puerto Rico already belonged to the United States. The invasion was pertaining to Puerto Rican lifestyle, and Santiago narrates the invasion through the eyes of a child, young Negi.

> Miss Jiménez [the new second and third grade teacher who came to teach the students English] came to Macún at the same time as the community center. She told us that starting the following week, we were all to go to the centro communal before school to get breakfast, provided by the Estado Libre Asociado, or Free-Associated State, which was the official name for Puerto Rico in the Estados Unidos . . . the Jun-ited Estates of America. Our parents . . . should come to a meeting that Saturday, where experts from San Juan and the Jun-ited Estates would teach our mothers all about proper nutrition and hygiene, so that

we would grow up as tall and strong as Dick, Jane, and Sally, the Americanitos in our primers. (Santiago 2006, 64)

First, English teachers are provided to begin the demotion of Spanish as the official language; second, free food is provided to attract the large lower class who is in need of food since they do not have jobs to buy food or land to produce their own food; and third, new ideals about raising children are necessary for children to be strong and to resemble American children.

Based on the descriptions of the provisions made by the Americans, Puerto Rican families are deemed to be in dire need of reform and America will help them with this reform free of cost. By attending the meeting, parents will learn how to take care of the children they have been providing for and protecting since their birth without America's help. What is even more amusing is the idea that once parents come to the meeting and receive the "proper" nutrition and hygiene directives, then their children can grow up to look like American children who are tall and strong—a new standard shrewdly imposed on the families by America. It is paradoxical how Americans have depleted the livelihood of the Puerto Ricans whereby Puerto Ricans were able to feed their children through farming the land for their own consumption, but America is now assuming the patriarchal role as the father of the nation who is in charge of the proper care of the Puerto Ricans. Whyte (2018) purports, settlers create moralizing narratives about why it is (or was) necessary to destroy other peoples (e.g., military or cultural inferiority)" (135). In this instance, America is creating a narrative about why their intervention in Puerto Rican family life is necessary—they do not have enough food for their children, their children are dirty, and they all need to learn English. To further advocate for the health of Puerto Rican children, Americans believe that the children look weak simply because they do not look like American children, since they are not eating American food. Americans do not consider that the genetic make-up of Puerto Ricans differ from mainstream white or even black Americans because of their predominant biracial or multiracial ancestry. America is clearly attacking the familial cultural practices of lower-class Puerto Ricans, planting the idea that their ways are inferior to those of America, and it is manifested in the physical appearance of Puerto Rican children.

Interestingly, Santiago does not write the children at Negi's school as politically uninformed about the colonial settlers of Puerto Rico. Through the character, Ignacio Sepúlveda, readers learn that parents are talking to their children about the political situation in Puerto Rico. Since the American invasion, in addition to free breakfast, free toothpaste and toothbrushes, and free basic food items for parents to take home like peanut butter and cornflakes that are not indigenous to Puerto Rican culture, there is now a nurse at the school ready to vaccinate the children. Ignacio tells Negi, "It's all because of politics. . . . My Papá says the government's doing all this stuff for us because it's an election year. . . . They give kids shots and free breakfast, stuff like that, so that our dads will vote for them [the American government]" (Santiago 2006, 71). Negi, unaware of this part of her history, goes home and questions her father. From him, she learns that Puerto Rico became a colony of the United States in 1898. However, a "lot of Puerto Ricans don't think that's right," says Papi, Negi's dad. He continues, "They call Americanos imperialists, which means they want to change our country and our culture to be like theirs." Negi then asks, "Is that why they teach us English in school, so we can speak like them?" Papi responds, "Yes." And Negi cunningly replies, "Well, I'm not going to learn English so I don't become American" (Santiago 2006, 73). Without knowing it, Negi is prepared to subvert the colonizer's power by refusing to learn their language. Her mother, also, unknowingly subverts American colonization as well by storing the American food they received on a high shelf, using them only when "hunger cramped [their] bellies" (Santiago 2006, 68). In effect, Mami decolonizes America's attempt to colonize their eating habits and food consumption and sets an example for her children to do the same.

The idea of settler colonization also provokes thoughts about taking the land and changing the people, which is explored by Santiago through the main character's family. As the discussion with her father ensues, Papi informs Negi that it will take more than resisting language domination to prevent herself from becoming American. He tells her, "Being American is not just a language, Negrita, it's a lot of other things. . . . Like the food you eat . . . the music you listen to . . . the things you believe in" (Santiago 2006, 73). As Negi absorbs this new information, she points out to her father how awkward Americans sound when they speak Spanish. Her father tells her that the ones who sound awkward are those "who don't take the trouble to learn it well." He states, "That's part of being an imperialist. They expect us to do things their way, even in our country" (Santiago 2006, 73)—an act that D. Ezra Miller (2016) refers to as "settlerscape" where the settler is inscribing or carving out their way of being onto the indigenous or colonized people. From this conversation, however, it is obvious that even the peasant class, which Negi belongs to, is aware of their history and the trickeries of the American government, and it will not be so easy for America to inscribe upon the Puerto Ricans' lives.

CONCLUSION: BEYOND AMERICAN INVASION

In the midst of American colonization, Puerto Ricans did not surrender their desire for independence. The more the United States increased its reach over the island, the more the Puerto Ricans fought to gain control over their land. In Blanca G. Silvestrini's (1989) "Contemporary Puerto Rico: A Society of Contrasts," she reveals that the United States had "strategic and economic interests on the island [that] were safeguarded by the acquisition of land for military purposes" (148). The United States was building its military base in the Caribbean to safe-keep its imperial power. Nonetheless, Puerto Ricans wanted a voice in their governmental affairs because US colonization denied them political autonomy, since the Olmstead Act of 1909 allowed the United States to have direct control over all Puerto Rican affairs.

An important tenet of the Puerto Ricans' struggle was their wish to continue their education in Spanish. However, the United States was opposed to this resistance because it wanted to Americanize the island and make the standard language English (Brás n.d.), a mandate of any settler colony. Between 1910 and 1920, students and teachers were arrested for striking against using English as the language of instruction in schools (Brás n.d.). Moreover, the persistence of many Puerto Rican nationalists who were fighting to have a voice in their country's affairs caused them to be imprisoned. Continued efforts to regain autonomy of their land caused Puerto Ricans to join in strikes and they created labor unions against the United States. Their efforts remained fruitless, until 1917, when under the Jones Act—the passing of which was led by Luis Muñoz Marín—Puerto Rico was declared a US territory and citizenship was granted to all Puerto Ricans; still, this did not mean more autonomy in their governmental affairs. At this juncture, they were citizens of the United States, and the president appointed all major Puerto Rican government officials, such as the governor, but there was no clarity about the political status of Puerto Rico, only that it was owned by the United States (Nicola Foote 2013, 201).

Economically, Puerto Rico continued to struggle. Silvestrini (1989) highlights a critical point in US history that sickled Puerto Rico's economy and led to mass migration to the United States. The Great Depression of the 1930s in American history greatly impacted Puerto Rico because Puerto Rico's economy was closely tied to that of the United States. Consequently, the island became even more dependent on the United States for survival during and after the 1930s. Silvestrini (1989) explains that the working-class families of Puerto Rico suffered the most because of the increased rates of unemployment on the island (151). It was during this time that Negi and her family migrated to America, but Negi was not happy about leaving her homeland. She states, "For me, the person I was becoming when we left was erased, and another one was created. The Puerto Rican jibara who longed for the green quiet of a tropical afternoon was to become a hybrid who would never forgive the uprooting" (Santiago 2006, 209). From 1940 to 1949, Silvestrini (1989) states that 151,000 Puerto Ricans migrated to the United States in search of a better quality of life (152).

Settler colonialism theory offers a dynamic lens through which one can situate the ongoing actions of colonizers on indigenous peoples. Indigenous peoples, as previously explained, are the people already living on a land, which settlers want to acquire. In this research, Puerto Rico has two sets of indigenous peoples—the Taìnos who were on the land when the Spaniards arrived, and the mélange of Spanish, Native Indian, and African populace who were on the land before the Americans arrived. Additionally, settler colony theory recognizes the environmental injustices committed against the land in the name of capitalism, imperialism, and expansion, and the anthropologic impact on a nation. It highlights the reorganization of the land for capitalist purposes, which are typically not organic, and which engender chemical degradation and pollution of landscapes. Ironically, settler colonials envision a reimagining of the land they believe will better serve the people than the traditional usage of the land by the indigenous peoples. In the long term, however, the land is abused, receives improper or no care, and organic biosystems are destroyed. Typical to every colonizer, though, a rhetoric must be created to justify their actions, and this rhetoric often surrounds the notion of the colonized/indigenous people as lazy and ignorant of how to make proper use of the land.

Intrinsic in any discussion of colonization, settler or otherwise, is the notion of forced migration as the indigenous people experience so much loss and displacement that they must find a new geographical space to thrive. The Caribbean is a known site of migration and Caribbean authors like Esmeralda Santiago, Junot Diaz, Edwidge Danticat, Julia Alvarez, and so many more are writing about their forced relocation due to settler colonization from the perspective of the colonized rather than the tainted narrative of the colonizer. In this way, the indigenous peoples are reclaiming power over the stories, their landscapes, and their culture without the element of the exotic or the Other.

Notably, while settler colonialism offers a useful ecocritical lens, what it does not adequately acknowledge is the colonized's ability to subvert colonial power. This is often ignored or marginalized in criticisms of the colonized, and it is a fact that needs to be underscored. Santiago's memoir provides distinct examples of this, illustrating that "while Patrick Wolfe's influential theorization of settler colonialism as a "structure not an event" remains useful, the critical focus on colonial power structures can marginalize the impact of decolonial resistance [such as Negi's and her mother's] and potentially inhibits the urgent work of rearticulating colonialism and imperialism" (Iyko Day et al. 2019, 8). In this way, there can be justice for the many acts of environmental injustice committed via settler colonialism.

BIBLIOGRAPHY

Ayala, César J. 1996. "The Decline of the Plantation Economy and the Puerto Rican Migration of the1950s." *Latino Studies Journal* 7, no. 1: 61–90. http://lcw.lehman.edu/lehman/depts/latinampuertorican/latinoweb/PuertoRico/ayalagration.pdf.

Blanca, G. Silvestrini. 1989. "Contemporary Puerto Rico: A Society of Contrasts." In *The Modern Caribbean*, edited by Franklin W. Knight and Colin A. Palmer, 147–167. Chapel Hill: University of North Carolina Press.

Brás, Marisabel. n.d. "Puerto Rico and the United States." *Library of Congress*. Accessed March 12, 2020. https://www.loc.gov/collections/puerto-rico-books-and-pamphlets/articles-and-essays/nineteenth-century-puerto-rico/puerto-rico-and-united-states/.

Day, Iyko, Juliana Hu Pegues, Melissa Phung, Dean Itsuji Saranillio, and Danika Medak Saltzman. 2019. "Settler Colonial Studies, Asian Diasporic Questions." *Verge: Studies in Global Asias* 5, no. 1: 1–45. doi: 10.5749/vergstudglobasia.5.1.0001.

Foote, Nicola. 2013. "U.S. Interventions and Influences in the Early Twentieth Century." In *The Caribbean History Reader*, edited by Nicola Foote, 199–202. New York: Routledge.

Higman, Barry W. 2011. *A Concise History of the Caribbean*. Cambridge: Cambridge Press.

LeFevre, Tate A. 2015. "Settler Colonialism." In *Oxford Bibliographies in Anthropology*, edited by John Jackson, 1–26. New York: Oxford University Press.

Little, Becky. 2017. "Puerto Rico's Complicated History with the United States." History.com. Last modified September 1, 2018. https://www.history.com/news/puerto-ricoscomplicated-history-with-the-unitedstates.

Lynch, Barbara Deutsch. 1996. "Caribbean Environmentalism: An Ambiguous Discourse." In *Creating the Countryside*, edited by DuPuis E. Melanie and Vandergeest Peter, 225–256. Philadelphia: Temple University Press. www.jstor.org/stable/j.ctt14bsstr.12.

Miller, D. Ezra. 2016. "'But It Is Nothing Except Woods:' Anabaptists, Ambitions, and a Northern Indiana Settlerscape, 1830–1841." In *Rooted and Grounded: Essays on Land and Christian Discipleship*, edited by Ryan D. Harker and Janeen Bertsche Johnson, 208–2017. Eugene: Pickwick Publications.

Santiago, Esmeralda. 1993. *When I Was Puerto Rican*. Reprint, Cambridge: De Capo. 2006.

Saranillio, Dean Itsuji. 2015. "Settler Colonialism." In *Native Studies Keywords*, edited by Teves Stephanie Nohelani, Smith Andrea, and Raheja Michelle H., 284–300. Tucson: University of Arizona Press. www.jstor.org/stable/j.ctt183gxzb.24.

Whalen, Carmen Theresa. 2005. Introduction to *The Puerto Rican Diaspora: Historical Perspectives*, edited by Víctor Vázquez-Hernández and Carmen Teresa Whalen, 1–42. Philadelphia: Temple University Press.

Whyte, Kyle. 2018. "Settler Colonialism, Ecology, and Environmental Injustice." *Environment and Society: Advances in Research* 9, no. 1: 125–144. https://kylewhyte.cal.msu.edu/wp-content/uploads/sites/12/2019/02/2018-WhyteSettler-Col-Ecology-EJ.pdf.

Chapter 11

"Coloniality" of Humans and the Ecology

An Ecocritical Reading of Shubhangi Swarup's Latitudes of Longing

Risha Baruah

For many years, mankind has lived through the Western project of political, economic, and cultural colonialism, which saw a shift of ideologies, narration, and knowledge during the twentieth century that witnessed the phase of decolonization. In consequence, counternarratives to the colonial discourse emerged under the efforts of several eminent thinkers like Edward Said, HomiBhaba, Gayatri Spivak, Franz Fanon, Chinua Achebe, and Ranjit Guha to name a few. Collectively, such literary and critical narratives have been regarded as a discourse to postcolonialism, which has essentially (counter) been narratives of the legacy of the Western Empire through the agency of imperialism, which seems to have outlived its period through the various forms of discrimination, categorization, stereotyping, and (mis)representation. The consequence of such efforts had been to fix the dualistic dichotomy of the Orient and Occident as opposing categories. In the recent years, there has been an increased output of literary and critical writings revolving around such notions, which had attempted to reshape the Western ideologies, interpretation, and cultural structures. Amid this, an advanced understanding of postcolonial narratives was discussed by Anibal Quijano under his foundational concept of "coloniality" of power, which was later developed by the Argentine semiotic Walter Mignolo. As a concept, it outgrew from the "Dependence Theory" and subsequently became a significant part of Latin American subaltern studies. For them, "coloniality" was not only an Eurocentric discourse but also an inseparable part of "modernity," which had been well illustrated by Walter Mignolo in many of his works, including his

2014 keynote lecture at the symposium of Guggenheim UBS MAP Global Art Initiative titled, "Coloniality: The Dark Side of Western Modernity" and Anibal Quijano in his article, "Coloniality of Power, Eurocentricism, and Latin America" (2000). Such positions attempt to expand the notion of colonialism from a political and social dominance of power to include the modern realities of consumerism, globalization, capitalism, and neocolonialism. This effort of "coloniality" aims to decentralize a singular narrative, which attempts to reduce all individuals and their experiences into "One Story" and thereby take away their variants in their identity and humanity. This idea was also vocalized by the famous Nigerian writer Chimamanda Ngozi Adichie in her TED talk 2009 under the title, "The Danger of a Single Story." In this context, coloniality attempts to explore multiple narrations within a "single story" so as to make literary efforts relevant and universal. In consequence, the concept has received more visibility and has subsequently assimilated to the varied cultural and literary contexts across the globe.

Following this trend, the theoretical world has witnessed the benefits of cross-disciplinary programs like the idea of "coloniality," which has been expanded with its close association with other mainstream approaches like feminism, classism, ecocriticism, Marxism, speciesism, and racism to cite a few. These efforts are a counternarrative of Western ideologies like gender, class, race, and anthropocentrism, which are essentially movements of resistance, equality, and liberal democracy. These developments have played an influential role in reshaping the boundaries of theories, interpretation, and ideologies to include newer challenges, situations, and vulnerabilities of the coming generation. Amid these radical changes, the latter part of the twentieth century to the present twenty-first century has been regarded as the period of postcolonial where counternarratives of the traditionally eroticized East have been dismantled. Along with these cultural shifts, the 1970s saw the emergence of "nature" as an important area of study with the realization of the "dying" earth followed by the 1990s declaration by the U.S. President as the "the decade of the environment" and the 1989 *Times* Magazine award to the person of the year bestowed to "The Endangered Earth" (Glotfelty 1996, xvi). This situation could be seen as a direct outcome of Western anthropocentrism, which saw its birth with the shift from animism and paganism to Christianity. Further, the Christian ideals gradually paved a way for cultivation, private property, and domestication of human life and the ecology. Subsequently, the humanistic tendencies of the anthropocentric project lead to "human-centeredness," which further saw an increased momentum through the following ages, most notably during the period of Renaissance, Enlightenment, Little Ice Age, and the Modern Age. Following this trend, the twenty-first century witnessed an intensified form of anthropocentrism through rapid and unplanned development, industrialization, world wars, capitalism, and the selfish actions of humans that have been blind to the rest of existence in their urge for superficial and temporary luxuries and endless demands. Such an attitude had paved the way to the Age of Anthropocene—Age of Humans. Anthropocene is often regarded as the period of history that has witnessed an ever increasing supremacy of humans over all creation, which was made possible with the advancements in science, technology, and the virtual space. This furthered the human "ego" and arrogance to attach intrinsic values to them while reducing all other ecological beings to instrumental value. This attitude justified man's exploitations and destructive actions toward the environment (Callicott and Frodeman 2008, 58). Such an act of negligence, misuse, and dominance of the ecological world by humans has frequently been regarded as dangerous and self-suicidal. In addition, William Rueckert claims that there is no such thing as an individual but only individual-in-context as "everything is connected to everything else" (1996, 108). This derogative treatment of the ecosystem has directly brought human life and sustenance under threat. This dire situation called in for urgency of awareness among humans, so that no further damage could be done to the ecology. Amid the overwhelming awareness of the importance of nature in the human life and its constant reoccurrence in literary writings, the need for an organized study of nature was recognized. In consequence, ecocriticism emerged as an intellectual approach whose main task was to study the relationship between humans and the physical ecological world as represented in literary writings (Glotfelty 1996, xvii). In this regard, ecocriticism maintains a multi-disciplinary affiliation, which makes it difficult to narrow down the approach into a singular definition. In fact, it has been an umbrella concept that was frequently associated with other significant theories of the contemporary period, which witnessed a "green turn" of the theoretical world. Such efforts have led to the growth of several sub-theories that are significantly growing to become independent approaches, namely ecofeminism, eco-ethics, animal studies, green studies, postcolonial environment, social ecology, and deep ecology—all of which attempt to challenge the canonical beliefs, so as to include unheard voices in literary narrations as well as reinterpretation of canonical works of the past. In this regard, ecocriticism has not been limited as a mere global approach but has expanded to become a planetary program that has not only been mind bending or mind blowing or mind boggling, but it has been mind expanding (Rueckert 1996, 108).

In this chapter, the ideas associated with ecocriticism and postcolonialism would be studied. However, the main objective of the chapter would be to make a comprehensive and detailed reading of both the theories as a merged concept. Such efforts have been undertaken by several eminent thinkers associated to ecocritical readings like Glotfelty Cheryll and Harold Fromm (1996), Graham Huggan and Helen Tiffin (2006), Upamanyu Pablo

Mukherjee (2010), Elizabeth DeLoughrey and George B. Handley (2011), Greg Garrad (2004, 2014) and Timonthy Clark (2015) to cite a few. However, the most significant effort of the merger of postcolonial and ecocriticism was seen with the publication of Lawrence Buell's *The Future of Environmental Criticism: Environmental Crisis and Literary Imagination* (2005), which marked the Second Wave of Ecocriticism. The work had played a crucial role in expanding the boundaries of ecocriticism from nature writings of American and British writings to include concerns from cultural, literary, and theoretical parameters. This new dimension in ecocriticism played an important role in filling gaps in the theory by drawing associations from different yet similar disciplines, which allowed in the expansion of the cultural discourse. The emergence of ecocriticism amid the postcolonial period had resulted in several efforts of reading the two distinct, yet parallel concepts under a singular lens. In this regard, the co-reading of postcolonial and ecocriticism appeared as a natural outgrowth. This new overlapping program essentially has been a forum of resistance to a singular dominance, (mis)representation, power, and authority, which has been controlled and manipulated by several social agencies and structures. Such efforts have been witnessed in several literary narratives across the globe and across different historical periods. In this chapter, we shall undertake a similar reading of postcolonial and ecocriticism in Shubhangi Swarup's debutant novel, *Latitudes of Longing* (2018), which has been a self experienced journey of the journalist-author for a period of seven years. It is a dense story of conflicting and changing situations across several generations in varied topography. It is not a singular narrative but typically "India" of multiple realities where nature and culture are in tensional relationship with the advent of modernity, technology, and globalization. These factors have subtly played an influential role in extending the rule and existence of the colonial Western power in the form of neocolonialism.

The most fundamental area of concern of postcolonial-ecocriticism has been the study of nature. Both postcolonialism and ecocriticism deals with the exploitation of the environment and its natural resources for the selfish ideologies and endless needs of humans. In both the theories, nature has become a site of exploitation, violence, manipulation, and destruction. Here, nature has not only been seen as a unique physical existence beyond humans but also as a cultural discourse, which has been repeatedly conceptualized and represented into a secondary and *lacking* category to normalize and justify the masked motives and purposes of humans. Frequently, ecocritics have associated the politics of dualism with the concept of nature alongside its marginalized and binary opposition with culture. Nature has been manipulated to the dualistic position of being permanent yet changing, calm yet chaotic, clear yet mysterious, powerful yet vulnerable, destructive yet self-nurturing, innocent yet dangerous. While at the same time, nature has been placed as a binary opposition to culture wherein nature has been regarded as passive, inferior, unpredictable, lacking, "natural," impulsive, unexplored, and exoticized as the "Orient" *Other*. As against it, culture has been regarded as temporary, detached, sophisticated, artificial, and polluted (Moore 2008, 55). A similar trend has been narrated in *Latitudes of Longing* through different generations, situations, and topographies. Swarup traces environment as diverse, blissful, abundant, fearless, and curative that has also been different, unknown, and difficult as it fades and bleaches "all things vibrant-colors, emotions, sounds, entanglements—have faded into their paler selves" (2018, 220). This immediately shows the complex fixation of nature as a stereotypical construct with whom humans share an ambivalent relationship. In addition, nature has been acknowledged with divine reverence, which we often exploit like the natives of the Andaman Island who "worshipped their gods" and ancestors through nature and were also "the ones they hunted" (Swarup 2018, 248). This shows that our association, conditioning, and manipulations of nature have been different at different periods and situations showcasing the fickleness, ambivalence, and the unethical position of mankind who have always selfishly sacrificed nature. It further highlights the deep-rooted anthropocentric attitude in us. However, it would be unjustified to read the densely layered narrative as a typical revisiting to the traditional ideas associated to ecocriticism. The novel has been a sincere attempt of the writer to show the endless possibilities, realities, and fluidity of the environment that cannot be translated but has to be experienced. Such a notion of reading the environment has been stated in the seminal essay of William Rueckert, "Literature and Ecology: An Experiment in Ecocriticism" (1978), where he regards ecocriticism as not just an interest in translating ecological concepts to the study of literature but also an attempt to see literature in the context of an ecological vision (1996, 115). Such an understanding of the natural world makes us realize that "one cannot judge the natural world by human laws" as it "doesn't adhere to the laws of science in the way scientists do" (Swarup 2018, 64, 310). This was because the ecosystem has been constantly changing to meet the daily battles for survival, which has made nature more adaptive and evolved. In this regard, Girija Prasad frequently claims that research studies appear inefficient, ignorant, and deluded as the mechanism and functioning of nature has been beyond the potentials of science and technology.

In addition to the complexities associated with the dualistic image of nature, the problem of (mis)representation has been intensified with the lack of language. While language has "usually been seen as a human endeavor" (Meijer 2016, 73), it has never been neutral. In fact, it has been politically driven by social ideologies and power structures of society. Ecocriticism has made attempts not only to speak about nature but also for nature, which has generally been regarded as a space of silence. However, such efforts have been made by man using "*his* language as the point of intersection between the human subject and what is to be known about nature," resulting in the misrepresentation of ecology in narratives and the biased justification of capitalistic attitude of humans (Manes 1996, 22).This has unfortunately pushed nature to a secondary space of marginalization. As a counter to this problem, Deep Ecologists have made attempts to listen to the nonhuman world and thereafter decentralize single narratives, which attempt to exclude presences that attempt to destabilize the normative discourses. For this, we must stop considering the silence of nature as its "curse." Instead, we should regard it as a "cue" for recovering a language appropriate to environmental ethics, more in tune with nature and outside culture (Manes 1996, 17). This idea has been deliberated by the elderly Apo to his new wife Ghazala at the end of the novel, where he says that "Lambs bleat. Dogs bark. Birds chirp. I snore, he says. It is the law of nature" when she complains of imaginary noises of the goat to her husband (Swarup 2018, 319). This incident immediately takes us to the beginning of the novel wherein a similar situation was encountered by the newly wed Girija Prasad and his wife Chanda Devi in the tropical warmth of the Andaman Island. This technique makes the narration circular and well connected with an organic plot despite the multitude of range in characters, situations, generations, histories, and terrains.

Through the earlier discussions in the chapter, it has been evident that nature has been positioned as the *Other*, a concept that has been closely associated with the theory of postcolonial. However, in recent times, it has become an important concept in several other relevant theories like feminism, psychoanalysis, Marxism, and post-structuralism. The "othering" of nature has played an influential role in the shifting of ecocentric ideologies to anthropocentric attitudes. Such assumptions can be traced through history, which makes it evident that humans were initially ingrained with the social beliefs of animism, primitivism, and paganism, which added the spiritual and divine aspect to nature that was revered and feared by humans. Such a relationship was maintained in the novel by the natives with the river Bagmati, which was worshipped by them as they feared calamities would befall them if they upset the spirits of nature (Swarup 2018, 276). For them, the natural world was a "living organism" whose spirits in the natural objects protected nature from man as well as man from nature. According to this philosophy, it was believed that the phenomenal world was alive in the sense of being "inspirited" not just by people, but also by animals, cultural artifacts, plants, and even "inert" entities (Manes 1996, 18). In this regard, the nonhuman world was alive and filled with articulate subjects that were able to communicate with humans. This not only encouraged in the protecting of the ecology but also in maintaining a relationship of coexistence between the two for a balanced harmony in the ecosystem. However, this social structure changed with the advent of Christianity, which associated divinity with mankind through the concepts of the Great Chain of Being and the Divine Rights of Kingship. It also paved way for cultivation, domestication of human life and the ecology. In this regard, Christianity can be considered as the most "anthropocentric religion," which slowly spread throughout the world with the Western project of colonialism (White 1996, 9). Such tendencies lead to the overlooking of sacredness in nature, which enabled in fostering an attitude of hostility and threat toward the environment. This colonial attitude has been frequently highlighted by Swarup in her novel as expressed in the words, "all they needed were tigers to shoot and the dream of the British Raj would be complete" (2018, 92). With the dominant spirit of anthropocentrism and individualism blending with capitalistic aspirations of the market, humans have subsequently drifted away from the concerns and situations of the natural ecosystem toward an increasingly privatized world of the "in-indoor." This intensified the tension between humans and the ecology, which was further aggravated by colonization as it intended to discover the unknown world to gain access for the exploitation, manipulations, and control over the natural resources of the non-Western world for the political and economic gains. This has been aptly addressed by Swarup in her novel where every aspect of the tropical Andaman Island was in balanced harmony with nature and "the only to the rule were the British" (2018, 15). In this regard, the reading of "nature" from the colonial perspective has been a spontaneous approach as it had distinctly shaped the discourse of nature. In fact, the pattern of "othering" of nature has been very similar to the process of postcolonial "othering" of the racial communities wherein the Occidental West has been placed as a superior category in the oppositional binary with the inferior non-Western world. In this dichotomy, culture has resisted the presence of nature while at the same time requiring its *presence* to acknowledge the superiority of the cultured world. This highlights the ongoing ambivalent relationship between humans and the ecology. Further, the postcolonial concept of Euro-centrism can be seen in the attitude of anthropocentrism, which attempts to place humans at the center of every dealings, policies, and actions of the ecosystem. In this attitude, the presence of humans becomes significant as the singular force that could alter the functioning of the natural world either in a positive or negative way. Such a stance makes us aware of the environmentalist attitude deeply designed in us that marks the advent of the Age of the Anthropocene. In such an influencing position, we attempt to improve and maintain the quality of the ecology but without any substantial efforts for radical changes that could hamper the comforts and luxuries of human beings. This highlights the shifting boundaries of ethics that has been frequently redefined so as to suit the purposes of a generation. For them, the ideals of science, technology, human rights, liberal democracy, and economic progress are in pivotal role even at the cost of environmental crisis (Garrad 2007, 19). Such a position emphasizes the hypocrisy of mankind that claim to have achieved enlightenment but has remained in darkness to the increasing concerns and challenges of the environment.

Through the narrative, we get a comprehensive insight that the colonial agenda was widely spread so as to expand the prowess of the Empire. In this context, the colonizers had advertised Andaman as a "free land waiting to be given to any community that cared to show up" thereby stripping off all its identity, culture, and dignity (Swarup 2018, 59). This was done to have an organized religion that could enlighten the "some uncivilized people" that lived on the island as the British Empire failed to cooperate with the natives due to the language and cultural barrier. Such efforts not only changed the original milieu of the island but also extended the impression of the British rule, which was evident in the description of the Goodenough's bungalow as the remains of the British colonial rule that was later inhabited by Chanda Devi and Girija Prasad who was referred as "the traitor who dresses and talks like the British" by the ghost of the Punjabi mutineer (Swarup 2018, 24). Here, it is important to take note of the process of naming environment, which could be seen as a device that reflects the ideals of coloniality, which extends its influence toward nature. In *Latitudes of Longing*, Lord Goodenough associates names like Breakfast Bay, Marmaladeganj, Parrot Island, Baconabad, and Crumpetpur, Mount Harriet, Ross Island, Rotten Eggs, and Spit Nest to the immediate environment that he explores in the island. In a similar trend, Daniel Hamilton in Amitav Ghosh's *The Hungry Tide* (2004) associated names like Lusibari, Canning, Jamespur, Annpur, and Emilybari to the newly explored ecological places in Sunderbans. In a similar fashion, natural objects in *Latitudes of* Longing were associated with names like the Alphonso mangoes, which later changed to Shivaji mangoes and likewise the Siachen Glacier was renamed as Kshirsagar Glacier Complex so as to mark the changing political identity of India. Such a tendency was seen in Bapsi Sidhwa's *Ice-Candy Man* (1988) wherein several characters changed their names and identities to dodge the communal and political riots, which were an immediate outcome of the partition and the Independence movement in India. While the influence of a political situation has limited to humans in the *Ice-Candy Man*, *Latitudes of Longing* shows that political and historical decisions, situations, and actions have had drastic implications that cross the threshold of human life to change the natural world. In fact, the entire process of naming the environment has nothing to do with history or our acknowledgment for nature. Instead, it has been an outcome of our immediate reaction to natural world, which is rarely based on the deep-reflection that nature deserves. In such a situation, the inferior Other (the non-Western and the ecological world) gets stranded to a marginal and secondary position with its origin, history, identity, and structures been stripped from them. Further, the naming process in both the narrations shows our insensitivity toward the environment, which we often categorize to suit our preferences, personal interests, conveniences, and relevance.

Such evidences of the human-environment relationship show us the advent of the Age of Anthropocene, which saw a new heightened role of human strength, influence, and responsibility limited not only to humans but also to the nonhuman world. It further regards "humans" as a new significant and its central role as the most decisive factor in the world (Raffnsoe 2016, xii, xiv). Such an anthropocentric ego furthered the "unscrupulous humans to crawl into, steal and profit from" nature and thereafter redesign the natural world to a new identity that is not indigenous and original (Swarup 2018, 62). This notion has been scientifically described by Swarup who narrates the historical introduction of the Gigantic Snail by the Japanese to the island, a native species of Malaysia and a source of high protein, which the Japanese introduced for military and survival strategies. However, with time, the species increased in number and became "one of the most destructive garden pests, second to the Barking Deer introduced by the British for game," threatening the balance of the native ecology (Swarup 2018, 17). In this regard, the "teak project" of Girija Prasad was also significant as it introduced foreign species to the archipelago for commercial profits. Despite the warnings from an old banyan tree through his wife Chanda Devi, Girija Prasad shows his confidence in science, research, and the intellect rather than the *real* opinions and signs of the environment whose voices always remain unheard. In fact, the tree said that although "teak was the most profitable timber in the world. If he could grow teak in the Andaman Islands, a perfect soil and climate for its growth, the Forest Department would be rich. But the banyan prophesied that teak would prove vulnerable to fungi the local flora had grown immune to" (Swarup 2018, 108). The predication was brushed away as illogical by Girija Prasad but it later became a reality. This shows that the ecological world knows its ways and roles in the larger ecosystem so as to maintain equilibrium of the environment, which has been generally broken by the artificial introduction of foreign agents for aesthetic, scientific, or commercial purposes by man. In consequence, these actions bring permanent ill-effects to the ecology by exposing new problems and challenges, which eventually make the environment vulnerable. In this new situation, humans feel more threatened, which further blinds us to acknowledge the qualities of the natural world. In this regard, we tend to associate negative connotations to the natural world like the description of the creek of the Middle Andaman as "a sinuous snake" where the soil was "enlightened" and atypical making nature behave ambitiously (Swarup 2018, 58). This immediately takes us to a similar description by Joseph Conrad in his famous work of *Heart of Darkness* (1902) where the

Congo River was described as an "immense snake uncoiled," which Marlow had to travel from London to meet the idealistic Kurtz (Conrad 2010, 10). Both the instances show the stereotyping of nature as a "dark" space that has often been reduced as a threat and a sinister. Such a notion of nature has been prevalent across cultures and history, which has reinforced negative associations to it leading to its eventual misrepresentation of ecology. Such conditionings have been maintained so as to hid our "lack" of understanding of the mysteries of nature and thereafter justify our malevolent actions.

With the increasing manipulations and control measures of nature, humans diabolically and ignorantly believe that they have the capacity to tame and exploit environment to their endless needs and selfish benefits. This reduces environment as a means to human ends. Since ancient times, humans have been using nature for survival and comforts as seen through the practice of hunting, which was predominately performed for food and protection. Such a necessity for hunting was later replaced to decorate the defeat on the hut's walls as "a gallery of animal skulls collected over decades," which was seen as a symbol of pride for the community (Swarup 2018, 174). However, this lifestyle and prioritization by humans have unfortunately pushed several nonhuman animals into extinction, which has resulted in the "receding paradise" that was created by the omniscient (Swarup 2018, 174). This greed and malicious behavior of humans have intensified through history that has often exploited nature for trading and economical purposes as seen in the case of Burma that was "blessed with all the precious gemstones and metals of the world. Amber, emeralds, jade, pearls, gold, platinum, even the world's biggest sapphires and rubies" were abundant, which was seen as a source of revenue by the colonizers who believed that "if they lose the mines, they lose the war" (Swarup 2018, 171–172). This shows the economical and political dependence of nations on the ecology to survive the complex politics of power, dominance, and authority. To achieve these ends, environment has often been manhandled and exchanged for capitalistic and political benefits as seen in case of the shrewd Calcutta businessman who was given the lease of the timber-rich jungles of the Andaman Islands by the government along with the exclusive felling rights in the island to export matchsticks in exchange for the first township, which he failed miserably in. In consequence, the angry political officials were only interested in taking legal actions against the businessman while completely ignoring the regenerative requirements of the altered environment. In a similar context, Plato and Thapa were found discussing a fragile political situation wherein they mentioned how the free access to weapons in the market along with wild spread of superstition has led to the increased killing of white elephants. In consequence, "their price has suddenly shot up. A single white elephant in the black-market costs more than the ivory from ten" while "opium is still the king" (Swarup 2018, 166).

Such practices highlight the deep-corruption, illegal smuggling, and uncontrolled greed for money that has driven man to threaten their own survival and existence by destabilizing the balanced ecosystem.

With the heightened influence of anthropocentrism, the modern period has witnessed an unprecedented increase in the interest for unplanned and uncontrolled development, which has resulted in compromising the environment. In this regard, the most common conversion has been the change of land pattern as natural spaces to overcrowded and busy cities with "people breeding like rodents" and "crawling like termites." In fact, the urban lifestyle has encroached the boundaries of the nature-centric communities with "cement structures fitted with proper toilets, ceiling fans, satellite televisions, satellite phones, sofas, gas stoves and other pollutants from an outside world" (Swarup 2018, 254). In this context, Thapa acknowledges the superficial shifts in our lifestyle that has created a trend of "fake" culture and glamour in the name of globalization, multiculturalism, and internationalism. In addition, he realizes that these changes have not only lead to an identity and cultural crisis but have also aggravated the modern man's dilemma of loneliness, emotional detachment, and alienation, which makes him feel like an "outsider" and the cities appears "all a distraction" (Swarup 2018, 250). Such a situation was reflected in the life of Devi, the wild and untamed daughter of Girija Prasad and Chanda Devi whose daily explorations and adventures remind us of Raka, a similar character in Anita Desai's *Fire on the Mountain* (1977). Both the children were rebels and curious who preferred the company of nature over their institutionalized schooling. In fact, Devi was called "Mowgli" by her father's brother for her matted brown hair and patchy skin who avoided school and clothes (Swarup 2018, 99). However, this situation changed when Devi was sent to a westernized boarding school in Nainital where she felt like an "outsider" as she was bullied, punished, and nicknamed as "Red Indian" for her burned skin and uncouth behavior in the boarding (Swarup 2018, 104). This shows the immediate process of *othering* of the unfamiliar presence and narrations in our cultural discourses. In this experience with culture, Devi developed a sense of difference, resistance, and hesitancy toward the natural world, which had earlier played a significant role in shaping her identity. In this new situation, she feels a "lack" throughout her life that does not allow her to be liberated like her younger self.

To worsen the conflicting situation in the novel, modernity paved way for the inclusion of agricultural machines and artificial intelligence in the snow desert, which otherwise was inhabited by a peaceful community. In this regard, the elder Apo of the community saw machines as a "monster" that squeaks the sound of war as well as "spewing dust for miles, contaminating their souls and fruits alike" (Swarup 2018, 262). The shift of ideologies from a nature-centric lifestyle to the new scientific and cultured society has been predominantly shaped by the westerns. Such changes accelerated other social challenges like the increasing inflow of global refugees and migrants that introduced poverty "like a weed" to the archipelago (Swarup 2018, 111). This idea of development has been aptly discussed by De Rivero as stated in the chapter "Development" by Graham Huggan and Helen Tiffin, which points it as a "little more than a myth propagated by the West that, under the guise of assisted modernization, re-establishes the very rift (social, political, economic) between First and Third Worlds that it claims to want to heal" (2010, 28). As understood, development had become a "self-privileging discourse of *neocolonialism*" that acted as an agency of extended imperialism through financial, technical, and military assistance. This new mechanism masks the exploitations by the west to the non-Western world as well as to nature, which together has been pushed to the category of the inferior and the uncivilized (Huggan and Tiffin 2010, 30).

The drastic manipulations and developments of the environment and human lives through extensive research and intellectual studies have reignited man's ego and arrogance to hold the superficial notion of his single-handed capacity to control the ecosystem. True to this understanding, Girija Prasad's natural inclinations were ruled by the ideals of science and intellectualism, which reduced his self-realizations to the predesigned cultural parameters. This was an outcome of his Western affiliations and connections in his academic and social life. Unlike him, his wife had embraced nature through her heart and spirit, which was evident as she connected with ghosts and also enjoyed "the laconic company of trees" (Swarup 2018, 5). Such a temperament had been an immediate consequence of her strong grounding in traditional living that was essentially nature-centric and spirited. In consequence, everyone considered her as a unique "goddess" that could control animals, ghosts, and her husband (Swarup 2018, 30). In a similar way, Apo had the capacity to understand the movements of the environment and thereby predict earthquakes and other natural calamities. He also considered "the tree as his friend. A dear and old one" as it was his loyal listener and his constant companion (Swarup 2018, 286). In fact, most of the characters in the novel have been directly or indirectly molded and influenced by their immediate environment. The most significant case has been the transformation of Girija Prasad after the death of his wife, followed by the marriage of his only daughter Devi and the eventual departure of the domestic helper Mary. In his loneliness and solitude, he finds himself evolving with nature by slowly understanding its mysteries through his immediate exploration of the island along his past observations of his deceased wife, which together reflected deep connections of remote and infinite elements of the ecosystem. In this situation of immediate exposure to the environment by himself, Girija Prasad slowly submits to the ecological forces of nature as "he no longer cares for himself—his British mannerism gives way to nature centric attitudes" (Swarup 2018, 115). At this point, he becomes a merged identity of culture and nature wherein he realizes that nature belongs neither to any individuals nor to any rules and boundaries of a nation. In fact, it is humans that belong to the larger ecosystem that we co-inhabit with other nonliving and living beings. These realizations in presence of the omniscient nature makes Girija Prasad feel "like an ant, shuffling around, tempted by the impossible" (Swarup 2018, 10). This truth of existence has also beenrealized by Apo in the snow desert, a world far away from the tropical Andaman, where he remarks that "even if the gods of machines and technology come down to help you. Apo pauses to think. Even if India, Pakistan, and China stop fighting over the ice and unite to remain there, the mountains will win" (Swarup 2018, 298).

The effort of this narrative was not simply to rethink the prescribed notions of nature and culture but also was to expand our realization that the environment has been constantly changing with our actions. In fact, this constant intrusion into the natural space has been simultaneously increasing tension between man and ecology, which Swarup attempts to explore through her novel. For this purpose, folktales and legends have been frequently mentioned throughout the narrative as they unearth hard-hitting lessons of truth, which calls for love and reverence toward nature, which could be hope for mankind. These also act as pointers of self-discovery that helps us to blur the fixed notions of society and thereafter encourage dialogues for the survival of the planet. In this regard, Wolfgang Sachs has introduced the concept of "global ecocracy" that attempts to see nature as an agency for living and survival and not simply as a market economy (Huggan and Tiffin 2010, 31). This subsequently encourages the ideals of sustainable living, which has become a significant voice for the contemporary eco-conscious generation that regards the ecological battle fundamentally important and worth fighting as without a planet, there would be no possibility for life and existence. Swarup's *Latitudes of Longing* charts multiple surreal, romantic, historical, and realistic adventures with a dense lyrical intensity in its prose following the fourfold terrains of nature across different culture and time periods to aptly show the extensive dominance of the "coloniality" praxis that has silently engulfed both humans and the ecological world to stand on logger-head with each other over the struggle for power, dominance, and survival through the mechanism of anthropocentrism. To counter this problem, Swarup advocates that ecology should be viewed from all aspects and perspectives as a central and normalized narrative, so that we could find possible ways for dealing with the problems of ecology. In this regard, the concerns of ecology are no longer limited to a local and global context but has expanded to become planetary issues, which if neglected can lead to the self-suicide of the planet and our civilization. From this understanding, the novel appears significant, especially in the time that we are living, which has become so automated, artificial, and virtual. In this context, *Latitudes of Longing* provides a refreshing and soul searching experience, which has been made possible by the eco-conscious journey through different environmental and emotional terrains that has been weaved by several generations of characters and situations creating a sense of "longing," "co-existence," and "exploration" of the readers and the fictional characters, their emotions, the environment tangled through past, present, and future filled with hope, which have been experienced at different degrees by most of the characters that are so inextricably woven into the fabric of the novel.

BIBLIOGRAPHY

Adichie, Chimamanda Ngozi. "The Danger of a Single Story." TED, October 8, 2009, YouTube video, http://youtu.be/D9lhs241zeg.

Armbruster, Karla, and Kathleen R. Wallace (eds). 2001. *Beyond Nature Writing: Expanding the Boundaries of Ecocriticism*. Virginia: Virginia University Press.

Buell, Lawrence. 2005. *The Future of Environmental Criticism: Environmental Crisis and Literary Imagination*. Oxford: Blackwell.

Callicott, J. Baird, and Robert Frodeman (eds). 2008. "Anthroprcentricism." In *Encyclopedia of Environmental Ethics and Philosophy*. New York: McMillan.

Clark, Timothy. 2011. *The Cambridge Introduction to Literature and the Environment*. New York: Cambridge University Press.

Conrad, Joseph. 2010. *Heart of Darkness*. New York: Oxford University Press.

DeLoughrey, Elizabeth, and George B. Handley (eds). 2011. *Postcolonial Ecologies: Literatures of the Environment*. New York: Oxford University Press.

Desai, Anita. 2012. *Fire on the Mountain*. New Delhi: Vintage.

Garrad, Greg. 2007. *Ecocriticism: The New Critical Idiom*. Oxfordshire: Routledge.

———. 2012. *Teaching Ecocriticism and Green Cultural Studies*. London: Palgrave.

———. (ed.). 2014. *The Oxford Handbook of Ecocriticism*. New Delhi: Oxford.

Ghosh, Amitav. 2013. *The Hungry Tide*. Noida: HarperCollins.

Glotfelty, Cheryll, and Harold Fromm (eds). 1996. *The Ecocriticism Reader: Landmarks in Literary Ecology*. Athens: The University of Georgia Press.

Hiltner, Ken (ed.). 2015. *Ecocriticism: An Essential Reader*. New York: Routledge.

Howarth, William. 1996. "Some Principles of Ecocriticism." In *The Ecocriticism Reader: Landmarks in Literary Ecology*, edited by Cheryll Glotfelty and Harold Fromm. Athens: The University of Georgia Press.

Huggan, Graham, and Helen Tiffin (eds). 2010. *Postcolonial Ecocriticism: Literature, Animals, Environment*. Oxon: Routledge.

Jackson, Mark (ed.). 2018. "Introduction: A Critical Bridging Exercise." In *Coloniality, Ontology, and the Question of the Posthuman*. New York: Routledge.

Love, Glen A. 2003. *Practical Ecocriticism: Literature, Biology and the Environment.* Virginia: Virginia University Press.
Manes, Christopher. 1996. "Nature and Silence." In *The Ecocriticism Reader: Landmarks in Literary Ecology*, edited by Cheryll Glotfelty and Harold Fromm. Athens: The University of Georgia Press.
Meijer, Eva. 2016. "Speaking with Animals: Philosophical Interspecies Investigations." In *Thinking About Animals in the Age of the Anthropocene*, edited by Morten Tønnessen, et al. Maryland: Lexington Books.
Mies, Maria, and Vandana Shiva. 2010. *Ecofeminism.* Jaipur: Rawat Publications.
Mignolo, Walter. "Symposium: Walter Mignolo on Coloniality and Western Modernity." Guggenheim Museum, December 6, 2016, Youtube video, http://youtu.be/ap3nqffXK3Y.
Moore, Byran L. 2008. *Ecology and Literature: Ecocentric Personification from Antiquity to the Twenty First Century.* New York: Palgrave Macmillan.
Mukherjee, Upamayu Pablo. 2010. *Postcolonial Environments: Nature, Culture and the Contemporary Indian Novel in English.* London: Palgrave Macmillan.
Quijano, Anibal. 2000. "Coloniality of Power, Eurocentricism, and Latin America." In *Nepantla: Views from South*, Volume 1, Issue 3, pp. 533–580. Durham: Duke University Press.
Raffnsoe, Sverre. 2016. "Introduction." In *Philosophy of the Anthropocene: The Human Turn.* London: Palgrave Macmillan.
Ruekert, William. 1996. "Literature and Ecology: An Experiment in Ecocriticism." In *The Ecocriticism Reader: Landmarks in Literary Ecology*, edited by Cheryll Glotfelty and Harold Fromm. Athens: The University of Georgia Press.
Said, Edward W. 2001. *Orientalism.* New York: Pantheon Books.
Sidhwa, Bapsi. 1989. *Ice-Candy Man.* New Delhi: Penguin Books.
Spivak, Gayatri. 2001. "Can the Subaltern Speak?" In *Norton Anthology of Literary Theory and Criticism*, edited by Vincent B. Leitch, et al. New York: W. W. Norton & Company.
Swarup, Shubhangi. 2018. *Latitudes of Longing.* Noida: HarperCollins.
———. "First Page: Sharin Bhatti Interviews Shubhangi Swarup, Author of Latitudes of Longing." Books on Toast, October 17, 2018, Youtube video, http://youtu.be/tjqG2FLjPtc. Wall, Derek (ed.). 2004. *Green History: A Reader in Environmental literature, Philosophy and Politics.* London: Routledge.
Westling, Louise (ed.). 2013. *The Cambridge Companion to Literature and the Environment.* New York: Cambridge University Press.
White, Lynn. 1996. "The Historical Roots of Our Ecologic Crisis." In *The Ecocriticism Reader: Landmarks in Literary Ecology*, edited by Cheryll Glotfelty and Harold Fromm. Athens: The University of Georgia Press.

Chapter 12

Women and Power
Digital Cameras in Postcolonial Caribbean Spaces in Literature

Denise M. Jarrett

POSTCOLONIZATION AND DIGITAL MEDIA IN CONTEMPORARY CARIBBEAN SPACE

Technology has taken over the world and has become the media for reporting, entertaining, and educating people about the vast cultures in different geospheres. Caribbean writers have begun to incorporate technology, such as digital media, not only as a symbol of modernity but, more importantly, also as a tool to present the different cultural practices, beliefs, environments, and peoples, creating a collage that allows readers to experience through different lenses and vicariously the diverse practices and peoples of the Caribbean Archipelago to promote a sense of empathy and to harbor a feeling of belonging. Because of the mutuality of colonial erosion of both the environment and the identity of the peoples, this telluric mass is of paramount importance to the peoples of the Caribbean islands. Postcolonialists and electronic media theorists coincide in contextualizing, debating, and decolonizing traditional perceptions of identity as affected by the environment. Postcolonialists have explored mimicry, stereotyping, exoticism, and primitivism as affecting identity development through subaltern theories, among others. However, green postcolonialism reveals the catastrophes of colonial interventions resulting in environmental atrophy, which affects the contemporary occupants. In awakening the consciousness about colonization and environmentalism, Frantz Fanon ([1963] 2004) elucidates the idea of a colonial city (66–67) and Edward Said ([1993] 2012) insists on an imagined geographical space that is associated with the politics of the land (326). In bridging this gap, green postcolonialism merges theories of ecocriticism and postcolonialism in search of justice. Thus, ecocritical postcolonial studies along with media theorists continue with the discussion on lucid ideas of virtual spaces for the creation of an identity in unjust societies.

In "The White Gyal with the Camera" by Kei Miller (2013) and "The Bonnaire Silk Cotton Tree" by Shani Mootoo (2017), the "virtual space" is created through a digital camera, which projects diverse, conflicting identities in postcolonial environments. Miller's "The White Gyal with the Camera" ("Gyal," Jamaican Creole, translates to "Girl" in English) delves into racism and power by examining the inner being of a seemingly heartless man who is mesmerized by the ghost of his forefathers' colonizers, a white woman with a camera, conflicting his identity because she captures the environment, and Mootoo's "The Bonnaire Silk Cotton Tree" divulges clashing cultural beliefs, classism, the destruction of the environment, and "Othering" that are affecting the protagonist and the marginalized people's identity in Trinidad through an Indian woman with a camera invading the environment. Within the framework of these two stories, media and postcolonial theories support the art of media as a virtual space that can affect and allow identity to oscillate in unjust postcolonial environments with racial, cultural, social, emotional, and economic conflicts.

SPACE AND POWER IN KEI MILLER'S "THE WHITE GAL WITH THE CAMERA"

While Kei Miller is renowned for his poems and novels, within *Kingston Noir* (2016), a collection of short stories edited by Colin Channer, featuring the geosphere of Kingston and its suburbs, appears Miller's short story, "The White Gyal with the Camera." Miller's point of reference is August Town, Eastern St Andrew, Jamaica, West Indies, a black metropolis, much like Southside Chicago, which he captures very frequently. In fact, his 2016 novel bears the name *Augustown*, a spin on the real township's name, "August Town." August Town is real; it is no mythical land like *Black Panther*'s Wakanda and likewise has no magical power via uranium that can sustain and elevate its people. Instead, it is a place where fairly recently, Paul H. Williams, a writer for Jamaica's premiere newspaper the *Gleaner*, quotes a Jamaican saying, " 'I will not be caught dead in that place,' 'Dat deh place dung deh soh, no sah,' 'It too volatile, no way' and 'Anything can happen, just so, and yuh dead' " (2015). In short, he believes that August Town is a death trap, a vast contrast to the pristine landscape of hills and valleys and of course, the miles of exotic white sand beaches in Jamaica.

With his focus on a settlement, it is evident that knowingly or unknowingly, Miller purports to architectural determinism—a theory that posits that human behavior is affected by the constructed or built environment. "Formerly called African Hill, Free Town and Pumpkin Hill, and now named after August, the month of Emancipation in 1838, [August Town] has evolved from a village settled by people who used to work on the Hope, Papine, and Mona [slave] plantations" (Williams 2015). To keep the descendant generations of former slaves subservient even today, poor infrastructure has led "to a throbbing inner-city community where crime and violence have compromised its historical significance" (Williams 2015). The introduction of a native son, Soft-Paw, the "don" from August Town (the same character in Miller's (2016) novel, *Augustown*), has created a cultural climate of fear that has imprisoned the gothic community especially during the dead of night. Like the suffering people who represent the fragments of once thriving slave plantations, the environment is on the periphery and is presented as almost diabolic. In this terrain, Soft-Paw tries to weave his way between personal and cultural misunderstandings and garners respect through violence to protect his masculinity and community but is finally moved by the picturesque sights of August Town in a white woman's camera. The story examines the inner being of a seemingly heartless man who is mesmerized by his environment via the ghost of his forefathers' colonizers, a white woman with a camera—"the white gyal with the camera."

August Town is a ghetto, and sociologist like William Bruce Cameron (1963) and several others purport that in most cases physical environment influences the behaviors of the people in that locale and can even define the activities that they engage in (Briney 2020). While Miller presents Soft-Paw as the don in the community and positions him as the authoritative figure who is feared by the residents and outsiders because of his unorthodox methods of justice or injustice, Soft-Paw becomes "soft" when he is confronted with the colonizer's power. Thus, he gives her a descriptive name that is ascribed as a symbol of authority. It is related that "she get the name because whatever Soft-Paw say we take it as gospel, and is Soft-Paw did send out word that if anybody see 'the white gyal with the camera' we was not to trouble her; we was to leave her alone" (Miller 2013). Miller uses a modern invention, a digital camera, as a tool in the hands of the colonizing force to mesmerize even the strongest anti-colonialist in the society with iconographies of the environment. However, the arrogance of the white colonizer is seen in her presumptuousness that makes her infiltrate a black society—August Town—because of her skin color without being ousted because she believes that she has the power to control. The courage and pompousness of the white woman is recounted in a way only the Jamaican dialect can describe:

> You had to give it to the white gyal though—is like she never have a coward bone in her body. She takes a plane to Jamaica and in my books that alone count

as bravery. Pretty blond girl on her own in the heart of Jamdown? Who ever hear of such a thing? But this white gyal take it further. Instead of staying at one of them hotels in New Kingston where she could order rum and Coke all day and listen to jazz in the gardens, or in a nice little apartment in Barbican or Liguanea, she did decide to rent a room right here in August Town. (Miller 2013)

It is not surprising that the "white gyal" could infiltrate the relic of a community that has turned its residents' control to "Soft-Paw," their protector and leader. It is within this environment that architectural determinism illustrates the creation of an impoverished town after slavery that still affects its current members who are descendants of former slaves. The former slaves were forced to the periphery on the steep rocky or sandy banks of a cascading riverside below their former plantation homes—the wealthy educated Mona Commons (now site of the University of the West Indies and former Mona Plantation) or the riches of Hope Pastures (previously, Hope Plantation), which is even more elevated geographically and socioeconomically than Mona. Stuck on the rocky hillside or sandy banks in August Town, with little to no infrastructure or independence, because the land is not conducive for farming, the people's behavior reflects the constrains that the society has placed on their lives; thus, the built environment affects their behaviors and attitudes.

The built environment has become a colonizing force to the people of August Town and as if to prove that this force exists, the ghost of the past colonizers appears in the form of a white woman. In introducing the woman as a colonizing agent in a society that has rendered the people of August Town victims of their environment, Miller (2013) uses technology to capture the beauty and vices of a desperate community through the lens of a white woman's camera. The name of the white colonizer seems to be irrelevant as her color is far more impressionable to even the man who is feared most by all in the community, the don Soft-Paw. Hence, the people garnered her name through the local newspaper. "It was when the papers come out with the gyal's picture print big and broad on the front page that August Town people did find out her rightful name. Marilyn Fairweather" (Miller 2013). August Town developed over time into a crime riddled ghetto because it has been neglected, and the people turn to crime to try and alleviate their suffering. After the former enslaved people were thrown off the plantations, they were displaced on lands that were not fertile or suitable for farming. Today, the fertile lands of the former plantations that once enslaved them to tend to their crops hold grandeur homes for Jamaica's rich and famous. On the contrary, the ghettos of Jamaica, one such is August Town, are known for violence, and strangers are not usually welcome, "but is like the white gyal with the camera never know or understand this—that she was living on grace—that if

Soft-Paw never send out such a word she would a dead from day one" (Miller 2013).

Miller's representation of Soft-Paw's donship shows how a society creates an environment through "complex relationships between individuals and their physical, social, and psychological world" (Marmot 2002, 252). August Town is supposed to be elevated as a historic center representing total rejection of colonialism, since the early, renowned community leader, Alexander Bedward, a great religious healer, is associated with the birth of Revivalism, an Afro-Caribbean religion, in Jamaica, and his church on August Town Road has been declared a historical national monument by the Jamaican government (Dennis 2015). Bedward became "leader to the Jamaica Native Baptist Free Church and founder of a movement that came to be known as Bedwardism, a powerful organization that catered to the spiritual, social, and physical needs of its followers" from 1894 to the early 1900s (Dennis 2015). Bedward is also credited as a strong anticolonial force because "many followers of Bedwardism became Garveyites and Rastafarians, bringing with them the experience of resisting the system, reaching community organization and self-reliance, and demanding changes of the colonial oppression and the white oppression" (Dennis 2015). Hence, "the story of August Town evolved around Bedward's great renown" (Williams 2015), but "the place that people would go to for healing became a place from which people would eventually flee to seek refuge elsewhere" (Williams 2015).

Ironically, over 100 years later, Miller creates the leader of August Town as the people's nemesis and a local terrorist to outsiders. Green postcolonialism is arguable as a defense for Soft-Paw's behavior. "In postcolonial discourse, justice is usually related to sovereignty and the ability of a people to maintain self-supporting, independent lifeways" (Roos and Hunt 2010, 3), which is what Soft-Paw tries to create in his environment. To "protect" the society, "Soft-Paw and the bwoy-dem was out there in the night, and to see them would make even a big man tremble, the way their trousers' pockets was big with guns" (Miller 2013). Soft-Paw's voice, walk, and actions speak of his cruelty to those who ignored his laws. To show the extremity of his actions, Miller (2013) allows the reader to see the contempt Soft-Paw has for the educated, whom he envisions as neocolonizers, when he attacks the university student. "Soft-Paw just flick out a knife and push the blade into the young man's back, not so deep that it could kill him, but deep enough" (Miller 2013). Through his violence, Soft-Paw has managed to keep out the police, who are referred to as Babylon in Miller's (2016) novel *Augustown* because of their oppressive nature. Soft-Paw seems to be in control to the point where he gives the "white gyal" the rules. "Do what you doing, but protection going to cost you. Hundred dollars a day. Hundred U.S. dollars. And a next thing: before you leave, you will have to show me all the pictures that you take" (Miller

2013). Reverse colonization is apparent as the "Don" now directs a race that once claimed ownership of his race. He even continues with a warning, reinforcing that he is the don. "Is me who run this place. You understand? Me is the community leader, and I don't want you take no picture that we wouldn't like. You get me?" (Miller 2013). Soft-Paw's action reflects an aspect of green postcolonialism where according to Roos and Hunt (2010), "the forms of justice—environmental, economic, and social—intersect" (4). Despite this bold assertion, Soft-Paw's power is nullified not only by the presence of the white woman but because of the vistas the lens of her camera unfolds.

The white gyal with the camera's boldness becomes the gravity that pulls down Soft-Paw from his elevated position of power to a man whose emotions are so overwhelming that he cannot control himself when he sees the panorama she has created. The images of his unsocial environment, August Town, reveal a beauty that captures Soft-Paw. Since the white gyal captures the town in the night, the narrator explains that "all day next day we was wondering so till we start to make joke that this so-called white gyal with her so-called camera must be some sort of vampire. What other kind of person would sleep during the entire day like she fraid of sun? Not a squeak nor a squawk from her during morning, midday, or afternoon" (Miller 2013). The vampire is an accurate description of the nature of colonialism, where the oppressors suck the life out of the oppressed. Thus, she starts sucking the power from Soft-Paw and the life out of the rest of the community.

Once she encounters Soft-Paw, his attitude changes. The white gyal with the camera orders him to look at the pictures without any recognition of his power. It must be her innate white privilege status that makes her think she could order even the vilest black man to look and he did! Not only did he look, but the emotions that these pictures evoke in Soft-Paw seem to be cathartic. Usually, "Soft-Paw face don't give away anything but I gather now that he was thinking he never before see August Town in the way that he was seeing it then—almost beautiful" (Miller 2013). The ghastly environment that he lives in becomes alive to him, and he sees the beauty even if it is in the gothic nature of the "owl, pale and bright on the roof of Miss Inez house" and "the old car that was rusting for years just at the end of the road" (Miller 2013). It is even more evident that he falls for the tactics of this woman who uses images to play on his masked emotions, so when "the white gyal with the camera [look] at him with a look that say he was almost beautiful too. He smile at her, his teeth brown as rust except for the one gold tooth glittering at the back" (Miller 2013). From this encounter, Soft-Paw becomes spongy in the palms of the colonizer. However, to remind the reader of the environment that has become the center of the community's problem and to show how Soft-Paw has been doped, the narrator communicates what the white gyal says:

"I think you have a really, how do you say, lovely place here." And she lift up her head and look all around and smile a smile that would make you think she was standing in the middle of fucking paradise—and mind you, Jamaica can be paradise when it want, like those times when you standing on a white beach looking at the moon sinking below the coconut trees. But this white gyal wasn't on no beach. She was in August Town. She was in the heart of the ghetto. (Miller 2013)

The narrator is showing how unaware the white gyal is of the people's plight in such an environment much like how the former colonizers presented the island as exotic and erotic even during slavery.

The community's trust wavered; then, they too were captivated by the white gyal who enthralled them. It was almost as if she hypnotized them to get their stories. She was capturing their culture and everyday lifestyles with her camera, hoping to harvest their stories in one or more ways. The white gyal's method of extortion was simple, becoming an active participant in the environment and culture to extrapolate the knowledge she wanted.

She go over to the table where them old fellows was playing dominoes . . . even though this white gyal had two things going against her—namely that she was white, and also that she was a gyal. A white gyal playing dominoes was even worse than a white gyal trying to shake her flat batty to Vybz Kartel or Beenie Man: them things wasn't normal; them things couldn't ever look right. (Miller 2013)

Although the people did not understand the white gyal's reasons for immersing in their culture, they let her in. The white gyal learns the art of infiltrating the black space to gain access as her forefathers did during colonialism that led to slavery and during slavery.

The charming views of August Town captivate Soft-Paw and the white gyal, causing them to act irrationally. Their irrational behaviors can be explained through a psychological dais by observing intrusive memories and images. According to Berntsen (2009) as mentioned by Brewin et al. (2010), visual "intrusions are instances of involuntary or direct, as opposed to voluntary, retrieval in that their appearance in consciousness is spontaneous rather than following a deliberate effort or search." The pictures seen through the white gyal's lens cause a spontaneous reaction, which makes Soft-Paw want to see the images over and over again. Brewin et al. (2010) suggest that "visual intrusions tend to be repetitive, uncontrollable, and distressing," which affect emotions and behaviors. There were "pictures that look like she was standing in the middle of the riverbed down by the part of town they call Angola [a name derived from the Angolan civil war 1975–2002 ("The Angolan" 2017)], the moonlight showing how the houses on the bank was close to falling in the sand" and "pictures from outside Judgment Yard, the red and green and yellow flags flapping in the night as if it was a balmyard and a cure for deep sickness was inside. And he see pictures of the actual balmyard—Bedward's church" (Miller 2013). After seeing pictures of the landscape and daily activities of the people, Soft-Paw expresses sadly that "he would love to always see August Town through the lens of the white gyal's camera, because he see things that he never see in all his twenty-nine years—a kind of loveliness in the people and in the place" (Miller 2013). Soft-Paw is unaware of the aesthetics of his space since he has created a gothic zone where he feels safe and does not necessarily want to expose his space to outsiders.

Despite the creation of a dark space, his love for the alluring visuals that Soft-Paw cannot see in the natural leads him to set up the stealing of the camera, while pretending that he is still the heartless don of August Town. The camera that is the white gyal's lifeline causes her to walk naked, a physical sign of psychosis, through August Town to confront Soft-Paw who has her camera and then slaps him in his face (Miller 2013). Soft-Paw's emotions after seeing his environment render him "soft" as he is doped by the same oppressive force as his forefathers.

Unfortunately, Soft-Paw accommodated the white woman in his environment where she has not only infiltrated the culture but the minds of the people, creating chaos even in a poverty-stricken unorthodox society with its own deadly rules and governance. The community is destabilized because its leader is weakened. What did the colonizers see when they entered Jamaica for the first time? A welcoming people who looked hostile and a beautiful untainted land that the Spanish colonizers thought was called *Xaymaca*—land abounding with springs (Atkinson 2006, 1); however, they entered fearlessly but eventually this meeting led to chaos when the natives tried to reclaim their lands. The natives who survived were then forced into spaces on the periphery where they could barely survive. Perhaps, then, Miller has adopted a microcosm of the history of Jamaica and has mirrored the story of the colonizer and the colonized in a modern scenario.

INDO-CARIBBEAN POWER AND SPACE IN SHANI MOOTOO'S "THE BONNAIRE SILK COTTON TREE"

In *Trinidad Noir*, Shani Mootoo's (2017) "The Bonnaire Silk Cotton Tree" also gives a paradoxical gothic grandeur of the landscape much like Miller's August Town, particularly the locations of the silk cotton trees in Trinidad and its green environment. The story opens with an environmental island legend:

At the beginning or the end—one decides as necessary which is which—of every village on the island stands a lone silk cotton tree. Woeful thou it appears, its naked trunk towers above neighboring trees. Above, its spread of branches houses birds, and at nights bays rest there. Below, among the cavities in its massive buttressed roots, live some of the island's largest snakes. It is home, too, it has long been known, to restless duppies and the mischievous yet irascible jumbie. (Mootoo 2017, 233)

The way to the silk cotton tree at Bonnaire was very dark (Mootoo 2017, 237), and "thousands of cicadas chirped and whistled, and crapauds croaked like dogs barking ... She stole through trees she knew by their smell—mostly guavas, but there were lime trees too and bush" (Mootoo 2017, 273), and after navigating the many plants, when she reached the clearing, in the moonlight, she encountered "an imposing silk cotton tree whose enormous and pale grey trunk stretched up like an extended arm, at the end of which were branches spread as if its mass were a hand opened wide to the sky" (Mootoo 2017, 238–239). It is in this space that Nandita Sharma, the artist, encounters the jumbie with her "heavy camera bag" across her chest and a tripod in her hand (Mootoo 2017, 28).

Mootoo, of East Indian descent, crosses cultural boundaries as she probes into Afro-Caribbean terrain since "the Moko Jumbie derives its name from West African tradition. The 'Moko' is an Orisha (God) of Retribution. The term 'Jumbie' was added post-slavery. The Moko Jumbie was regarded as a protector whose towering height made it easier to see evil before ordinary men" ("Traditional Mas" 2018). Hence, the jumbie manifests its traditional role as a protector who seeks retribution for the disregarded people and the erosion of the environment. Green postcolonialism in its effort to promote justice can be used to explain the stance of the jumbie as environmental justice ecocriticism, which "tends to value conservation/preservation" (Roos and Hunt 2010, 3). Within this space, Nandita, Indo-Caribbean, of affluent background, finds herself powerless with her camera, and the "Other," the jumbie and the duppies, give voice to protect the environment and the marginalized people of Trinidad.

Nandita seeks to find power and autonomy with her art because she did not want to be dependent on her father's wealth. In exhibiting fifteen pieces of art on seafood, Nandita has not captivated the audience because according to McLuhan ([1964] 2011) ideology, "people adapt to their environment through a certain balance or ratio of the senses, and the primary medium of the age brings out a particular sense ratio, thereby affecting perception" (337). The digital photograph from the video camera has become a primary medium, so still photography no longer appeals to the people unless the space is emblematic of their experiences, and certainly fifteen pictures of seafood do not appeal to even the person with a grand gustatory perception. Media theory, then, bemoans that, "a medium is not simply a newspaper, the Internet, a digital camera and so forth. Rather, it is the symbolic environment of any communication act" (Gupta 2006, 32)

With the perception that her work can become a symbol of the environment, particularly one that is taboo in her Indian culture yet a prominent part of Trinidad's folk culture, Nandita thinks about appealing to her audience. She feels that she has the power to exploit the environment. Hence, Father O'Leary influences Nandita who uses his newspaper column as an altercasting medium because as the sender of the message, he stirs a seemingly accepted counter-role for the "other" (Pratkanis 2011, 32) since he frequently expounds on "the dark arts of the Caribbean" (Mootoo 2017, 233) in trying to Christianize the people as the colonizers did because they did not want to understand the cultures of the people as one of their main aim was to overpower them. Nandita, a Hindu, like others is intrigued by the "sinful" stories written by Father O'Leary about those who visit the silk cotton tree, decides to follow his lead to Bonnaire to use her camera to capture the essence of the silk cotton tree and its inhabitants, along with any visitor that might fearlessly pay a visit. Father O'Leary denotes the uselessness of the silk cotton tree as he promotes it as an impractical piece of nature by edging on the colonial rhetoric. "Silk cotton trees serve no purpose today . . . . Before civilization, the Caribs and the Arawaks use to chop down silk cottons. They use the trunk of a single tree to carve out a whole canoe" (Mootoo 2017, 236). Thus, it is not surprising that nature is being depleted since, like O'Leary, there is no respect for the natural environment as habitat for some inhabitants of the earth. This idiosyncrasy is in tune with Graham Huggan's and Helen Tiffin's (2010) postcolonial environment ethics that pits Eurocentric colonial ethics against green postcolonialism (23) by juxtaposing "the conquest of nature [as] all part and parcel" (Roos and Hunt 2010, 3) of colonial conquest.

Nandita's act is akin to the white gyal with the camera as they both enter feared spaces without any regard for the malevolent reputation of the inhabitants or the environment. Furthermore, it is quite interesting that both Miller (2013) and Mootoo (2017) reverse the roles of the powerful in Caribbean society by giving women power through technology. Greg Satell (2014) in his article, "How Technology Is Creating New Sources of Power" argues that "today, information itself and the capability to process it has become a new source of power." While the power of these women subverts the male dominance in Caribbean societies, it brings to the forefront that women are also becoming more technologically advanced; thus, they try to manipulate societal power structure, and in doing so, in both stories, they use the environment.

In the clearing, the feared periphery, where she encounters the "imposing silk cotton tree" (Mootoo 2017, 238), Nandita becomes dramaturgical with the folklore character, conversing with the jumbie, who along with the duppies demand a space on camera to protest the depletion of the environment—the forest and their home and the criminal activities that have left many dead in Trinidad. While Miller (2013) promotes the heart of the don in Jamaica, if there is ever a don in Mootoo's (2017) story set in Trinidad, it would be the jumbie. The jumbie, like Soft-Paw defends his territory, but differently, explaining that "we live with the sound of quarrying . . . Bulldozing all the trees and all the bush, all the flowers, all the nests, the homes of all the animals and of all of we" (Mootoo 2017, 241), which reflects both the destruction of the natural landscape and gentrification of the marginalized space in the country. The jumbie calls for the education of the people about deforestation; "O'Leary free advertising don't mean nutten if the forest disappear. Why he don't use his big words to educate people, to tell them that the forest disappearing right before they eyes" (Mootoo 2017, 341). The jumbie and the duppies are cast as the "Other" as they reflect the concerns of the marginalized in Trinidad. Gayatri Spivak (1994, 1996) shows how the colonizers have marginalized and rendered their subjects voiceless. "Can the Subaltern Speak?" and "Poststructuralism, Marginality, Postcoloniality and Value" explore the "subaltern," one who lives inside a space and has great knowledge of that space, which is not influenced by an outside power.

Nandita is infiltrating that space much like how the white gyal infiltrates August Town; however, the jumbie will not allow the oppressor to tell his story which is in league with Spivak's thoughts as she contends that the "subaltern" is the most appropriate to impart knowledge about his or her culture and people. Thus, the jumbie is aware of his space; hence, he cries out for justice for the marginalized. "You see all the killing in the country?" "People getting killed like flies. All this casual brutality." "The defenseless dead, the unfought for dead, they restless for so" "They all right here now. They ent going to have no peace until their killers are caught and bought to justice" (Mootoo 2017, 245). Like Spivak, Edward Said (1993) in *Culture and Imperialism* depicts the colonized as the displaced "Other," having no voice or identity in the colonial space. So, Nandita is rendered powerless with her camera by the subaltern as the jumbie starts "gatekeeping," a theory by media theorist David Manning White from1950 but is still relevant today, stating that "the gatekeeper decides which information will go forward, and which will not. In other words, a gatekeeper in a social system decides which of a certain commodity—materials, goods, and information—may enter the system" (Green 2015, 56). Although, Nandita thought that she was in charge of her camera and what she produces as an artist, the jumbie forces her to agree to repay for his photoshoot by exposing him and the duppies as they protest not only the physical depletion of the environment but also the murders from all ages in Trinidad including during colonialization and slavery.

> Every person who get killed on the island since the beginning of the first injustice all the way to the present day wantonness—from the native people in the days of the early Spaniards, to the slaves of the British, to the present day victims of robberies and drug related and poverty related greed related envy, and jealousy, and power related crimes, all the people whose murders weren't caught. All of us on whom justice turn its back. (Mootoo 2017, 240)

The jumbie with the duppies have decided to expose the treachery of the colonized space by protesting during Carnival in a novel way as masqueraders, which is not an unusual feat. Moko Jumbies who are traditionally masquerade characters usually, "make long strides balanced on stilts that can be from 10 to 15 feet in height" ("Traditional Mas" 2018) because "the Moko Jumbie was regarded as a protector whose towering height made it easier to see evil before ordinary men" ("Traditional Mas" 2018). In masquerading, the jumbie and the duppies will take the saying "play yuhself" (play yourself) literally. This Trinidadian expression is explained as:

> In a culture of multiple identities, to play yuhself, as the Carnival cliché suggests, may be best achieved in fantasies of the so-called "other." This reality complicates our understanding of carnival masking in a situation in which "dressing up" and "dressing down" in the carnival season means more than temporary release or momentary social inversion. (Riggio 2004, 94)

Despite the fact that the jumbie and duppies will literally be playing themselves as masqueraders, they will still use Carnival as Riggio (2004) expresses as a vehicle for "release or momentary social inversion" (94) since the occasion will allow them to make a social commentary within the act of protest. Moreover, carnival, with its masquerading, is a form of social satire and is also seen as a form of protest (Mason 1998, 9; Riggio 2004, 256). The promise is that while all the people with digital cameras will capture the event, only Nandita will be able to produce pictures of the jumbie and the duppies. Thus, in paying the jumbie, a must, according to Father O'Leary, Nandita will become a renowned artist.

Additionally, Mootoo (2017) brings to the forefront the antiquated colonial belief that the artist is not productive and that art is not a profession. Nandita deemed herself an artist and moved out of her father's house, as an act of rebellion against her father's colonial regime. As an Indian man and a capitalist, owning the largest newspaper company in the island (Mootoo 2017, 235), the father's neocolonialist attitude leads to her rebellious spirit averting her conformation to colonial and cultural expectations, in which there is the stereotypical belief that an East Indian should study medicine, engineering, computing, any other science, or at least business. Therefore, when she fails at her third attempt at exhibiting her photographs, and her father remarks, in his toasts to her that this is "the conclusion of that phase of [her] life, and the hope that another is about to begin" (Mootoo 2017, 235), she decides to venture out. In addition, after being heavily criticized by the newspaper reviewer, she seeks a way to prove her craft as worthy.

It is only when she encounters the jumbie that she realizes that neither her class nor her camera is powerful in that space, depicting colonialization in reverse because she becomes colonized by the jumbie. Mootoo also reflects on the East Indian's traditional position as the "Other" in British colonial Trinidad as indentured workers, who also suffered under the colonizer's harsh measures. There is also the sense that Mootoo looks at the contrast where many East Indians in Trinidad are positioned as the "new" colonizers by many Afro-Caribbeans because of their economic status and their hunt for positions in high places and, most importantly, their support and participation in depleting the environment because many of them have been developing natural landscapes into industrial and upscale housing complexes. Roos and Hunt (2010) in presenting green postcolonization purports:

> Similarly, such ecocriticism recognizes that subtexts of class and race inhere in notions of nature or wilderness, that when put into practice, environmentalism often inadvertently results in racist or classist policy. In these ways, environmental justice has moved ecocriticism to consider how disenfranchised or impoverished populations the world over face particular environmental problems. (3)

However, the Indian, Nandita is rendered powerless just like Soft-Paw who becomes debased in spaces where they are known to wield power.

CONCLUSION: THE ENVIRONMENT AFFECTING WOMEN WITH DIGITAL CAMERAS AND POWER IN CARIBBEAN SPACES

The environment and colonization become of paramount importance in the stories. While the two women, white and Indian, enter spaces where they thought they were able to control with their cameras, they soon learn that their technology did not give them full access and authority in these spaces. However, they tried to exploit human spaces, which are peoples' or creatures' living environment. These exiled spaces are left to deteriorate or to be depleted with little concern that they are people's homes, and these people will suffer if their environments change. Postcolonial theories have explained some of the variances when it comes to the treatment and preservation of one's environment, but due to colonization and its remnants, people in these societies still feel threatened by outsiders who are not concerned about their suffering. It also shows that women are weakened despite their boldness, which can be read negatively by feminists, but the women symbolize the weaknesses of the colonial powers akin to the stereotypical colonial presentation of women as weak—the "white gyal" becomes mad, and Nandita is doped. With this outcome, both Miller (2003) and Mootoo (2017) emasculate the colonizers' and colonialists' ideals in the Caribbean because women of these races are seen as superior to Afro-Caribbean women. Furthermore, architectural determinism supports the art of resisting even through violence because it blames the society for the people living in derelict conditions. Additionally, while the don, Soft-Paw, holds the power in the community and subverts the oppressive class, he has pseudo-supremacy because of affective gratification when he encounters the ghost of the colonizer with her digital camera. On the contrary, the jumbie subverts the oppressive class as he uses technology to his advantage much like how Nandita wants to use technology as a tool to gain power. Mootoo uses someone from the Indo-Caribbean race to remind readers that Trinidad's cultural beliefs are not monocultural, so an Afro-Caribbean mythology becomes a part of the Indo-Caribbean world. Consequently, Miller (2003) and Mootoo (2017) suggest that the colonizer and the colonized are presented in unstable spaces, and colonial attacks on the environment are impacted by technology, in this case, the digital camera.

BIBLIOGRAPHY

Atkinson, Leslie-Gail. 2006. *The Earliest Inhabitants: The Dynamics of the Jamaican Tainos*. Kingston: University of the West Indies Press.

Black Panther. 2018. Directed by Ryan Coogler, Marvel Studios.

Brewin, Chris R., et al. 2010. "Intrusive Images in Psychological Disorders: Characteristics, Neural Mechanisms, and Treatment Implications." *Psychological Review* 117(1): 210–232. doi: 10.1037/a0018113.

Briney, Amanda. 2020. "What Is Environmental Determinism?" *ThoughtCo*. https://www.thoughtco.com/environmental-determinism-and-geography-1434499.

Cameron, William Bruce. 1963. *Informal Sociology: A Casual Introduction to Sociological Thinking*. New York: Random House.

Channer, Colin. 2012. *Kingston Noir*. Boston: Akashic Books.

Dennis, Denise. 2015. *Preserving Bedward's Legacy*. Jamaica Information Service, 31 July 2015. http://www.jis.gov.jm/preserving-bedwards-legacy/.

Fanon, Frantz. 2004. *The Wretched of the Earth*. New York: Grove.

Green, David. 2015. *Unfit to Be a Slave: A Guide to Adult Education for Liberation*. Boston: Sense Publishers.
Gupta, Om. 2006. *Encyclopaedia of Journalism and Mass Communication Volume 1*. Delhi: Isha, Books.
Huggan, Graham, and Helen Tiffin. 2010. *Postcolonial Ecocriticism: Literature, Animals, and Environment*. New York: Routledge.
Marmot, Alexi. 2002. "Architectural Determinism: Does Change Behavior?" *The British Journal of General Practice* 52, no. 476: 252–253.
Mason, Peter. 1998. *Bacchanal!: The Carnival Culture of Trinidad*. Philadelphia: Temple University Press.
McLuhan, M. 2016. *Understanding Media: The Extentions of Men*. McLuhan: McGraw-Hill.
Miller, Kei. 2013. "The White Gyal with the Camera." PDF. http://www.akashicbooks.com/the-white-gyal-with-the-camera-by-kei-miller-from-kingston-noir/.
———. 2016. *Augustown*. East Rutherford: Penguin.
Mootoo, Shani. 2017. "The Bonaire Silk Cotton Tree." In *Trinidad Noir: The Classics*, edited by Earl Lovelace and Robert Antoni, 233–249. Boston: Akashic Books.
National Carnival Commission of Trinidad and Tobago. 2018. "Traditional Mas Characters – Moko Jumbie." https://www.ncctt.org/new/index.php/about-ncc/departments/regional/ trad-carnival-characters/337-traditional-mas-characters-moko-jumbie.html.
Pratkanis, Anthony R. 2011. *The Science of Social Influence: Advances and Future Progress (Frontiers of Social Psychology)*. New York: Psychology Press.
Riggio, Milla Cozart. 2004. *Carnival: Culture in Action: The Trinidad Experience*. New York: Routledge.
Roos, Bonnie, and Hunt Alex. 2010. *Postcolonial Green: Environmental Politics and World Narratives*. Charlottesville: University of Virginia Press.
Said, Edward. 2012. *Culture and Imperialism*. New York: Vintage.
Satell, Greg. 2014. "How Technology Is Creating New Sources of Power." *Forbes*, October 2014. https://www.forbes.com/sites/gregsatell/2014/10/19/how-technology-is-creating-new-sources-of-power/#4ff4f3ae24ed.
South African History Online. 2017. "The Angolan Civil War" (1975–2002): A Brief History." https://www.sahistory.org.za/article/angolan-civil-war-1975-2002-brief-history.
Spivak, Gayatri. 1994. "Can the Subaltern Speak?" In *Colonial Discourse and Post-Colonial Theory: A Reader*, edited by Patrick William and Laura Chrisman, 66–111. Columbia: Columbia University Press.
———. 1996. "Poststructuralism, Marginality, Postcoloniality, and Value." In *Contemporary Postcolonial Theory: A Reader*, edited by Padmini Mongia. Huntington Beach: Arnold.
Williams, Paul H. 2015. "Dreams of an August Town Rebirth." *The Gleaner*, 7 August 2015. https://www.jamaica-gleaner.com/article/news/20150807/dreams-august-town-rebirth.

Chapter 13

Nature and Resistance in Coetzee and Abani

The Transcoporeal in African Fiction

Puja Sen Majumdar

The reason for bringing together J. M. Coetzee's *Life and Times of Michael K.* (1983) and Chris Abani's *Song for Night* (2007) is that: they are both stories concerning planetary entanglement, an entanglement, which is deeply connected with marginalization and vulnerability. In both narratives, the idea of the specter is used in extremely interesting ways where the central characters can interact with nature and land, in certain cases with the spirits of their ancestors as well and yet they are largely rendered invisible by the state. Both the writers decentralize the human and situate them as an entity, which exists within a network of other entities: animals, plants, nature, rivers, and water-bodies, with the knowledge that human life is not just interconnected but is dependent on cooperation with nature. The question that comes up here is that, can these characters be ever looked at as sovereign subjects? Or are they merely disposable bodies that can be exploited for labor in the same way that the lands and rivers that they are connected to are exploited by imperialism? It can be argued that even if they seem to be nothing but "bare life," the authors are able to unveil the agency of these characters through narrative techniques that must be explored. Following Giorgio Agamben, one can argue that the central characters of both these works are people oppressed in such a way that their existence (physiological, digestive, psychological, etc.) is completely controlled by the state and its repressive machineries and yet both of them become figures of resistances because of the connection that they share with nature. In fact, nature does not only become a space of refuge for these characters but also a resistant entity in itself against colonial powers. In that sense, maybe these texts can offer us new ways of thinking the subject as well as new ways of thinking about biopolitics and necropolitics.

The binaries of the human and the nonhuman, the passive and the resistant are broken down in the process and at the same time, death of a human body is not seen as the end of life but only as change of form. This chapter will try to read these two works in conjunction with theories of animism and trans-corporeality to see how the protagonists only have agency because they belong to an entangled world, a world where Michael K. can find his mother and the women before her through his relationship with the land and a world where sacred lakes and forests possess great power because they are repositories of human and animal souls. This chapter is divided into three sections. The first section discusses Coetzee, the second section discusses Abani and the third section explores in detail the possibilities of eco-critical resistance in both the texts.

J. M. COETZEE, BARE LIFE AND IMPOSSIBLE COMMUNITIES

The story of Michael K. is set in South Africa in the apartheid era (1970–1980s). It focuses on a fictional civil war during which the central character journeys from Cape Town to his mother's birthplace (Prince Albert) while trying to avoid camps, where people are reduced to vagabonds, dehumanized constantly and exploited for free labor. The entire novel is an exploration of how Michael K. manages to escape these labor camps. Michael K. is seen as a "deformed" person, both physically because of his cleft lip and psychologically because he does not seem to understand the warnings and threats extended to him by state machinery. The marginality of Michael K. becomes clear at the very beginning of the novel, as Coetzee describes the character's childhood by saying that he grew up sitting on a blanket looking at his mother polishing other people's floors and learning to be quiet. By learning to be quiet, Michael also learns how to be unnoticeable to his advantage and how to escape. And till the very end of the novel, he manages to run away every time he is incarcerated. However, the problem that Nadine Gordimer raises with this novel is whether Michael K's private destiny can be translated into any political resistance. Keeping in mind the complexities of the problem, one also has to remember that throughout the novel, there are several instances of solidarity and empathy between Michael and other oppressed men and women. At the same time, what is truly significant in the novel is perhaps his solidarity with the land that he wants to belong to. Even though it looks like Michael is attempting to escape history, history is also not the only way one can remember what has gone by. Michael does not know who his father is but that has a different legacy. A re-ancestralization happens when Michael returns to his mother's land. When Michael imagines a child in the house in Prince Albert, he can also see behind that child, in the doorway, her face obscured by shadow, a second woman, the woman from whom his mother had come into the world. And he thinks:

> When my mother was dying in hospital . . . when she knew her end was coming, it was not me she looked to but someone who stood behind me: her mother or the ghost of her mother. To me she was a woman but to herself she was a child calling to her mother to hold her hand and help her. And her own mother, in the secret life we do not see, was a child too. I come from a line of children without end. (Coetzee 2015, 131)

He keeps trying to find a child who is alone without a woman standing behind and fails to do so. At the same time, Michael desperately attempts to resist classification and categorization by the state. In the novel, his body is marked in other, more private, ways. As Jane Poyner elaborates, Michael K. "authors his own body, a body that colonial and apartheid discourse have sought to inscribe, define and regulate" (Poyner 2009, 88). This is a recurring trope in several of Coetzee's novels. If one looks at *Foe* and *Waiting for the Barbarians* as well, liminal bodies always find a way to escape control. In these two novels, it becomes difficult to interpret certain actions of Friday and the "barbarian" girl, respectively. It is interesting to note that both these characters have visible "deformities." The girl in *Waiting for the Barbarians* cannot see properly in one eye because of brutal torcher. Friday's tongue has been cut out by his colonizers. Thus, institutional oppression prevents them from expressing themselves at several points in the novels. The white characters are at times desperate to understand what they are thinking and fail to do so. This resistance to interpretation both by privileged characters within the narratives as well as the readers haunts Coetzee's texts. Even in this novel, such a resistance takes center stage. Michael feels a gap or hole or lack within himself but even then, there are some truths about Michael that ring truer than the others. At first, the medical officer thinks of him as a figure of fun. Michael is obedient to the authorities in the sense that when they ask him to jump, he jumps, and he keeps doing it until he falls. However, Michael rejects the food that he is offered by the state. The medical officer eventually confesses:

> Then as I watched you day after day I slowly began to understand the truth: that you were crying secretly . . . for a different kind of food, and only that. Now I had been taught that the body contains no ambivalence. The body, I had been taught, wants only to live. Suicide, I had understood, is an act not of the body against itself but of the will against the body. Yet here I beheld a body that was going to die rather than change its nature. (Coetzee 2015, 180)

Toward the end of the novel one also gets glimpses of what Michael himself thinks about his journey. He tells himself that he was mute and stupid in the beginning and he remains mute and stupid in the end. But now, there is a difference since he accepts his role as that of a gardener. It is important to take into account that instead of thinking of himself as human he calls himself an earthworm, an earthworm who knows that his job is to plant seeds using all the time he has. Jane Poyner also points out that by doing this "Michael K sets about writing the land, not only through the inversion of Crusoe's conquest of time and space, but also through the inversion of the ideology underpinning the Afrikaner plaasroman or farm novel" (Poyner 2009, 84), since in the plaasroman, there is always an implied claim on land. The idea of mute, stupid, or simple has been redefined by Coetzee and the notion of how the "subaltern" can speak has been complicated. Michael, the narrator claims, understands that there is nothing wrong with being simple since a simple person such as him can dream of planting seeds, which people who dream of constructing more and more camps do not understand. The world that is described at the end of the novel is a world where

> they were locking up simpletons before they locked up anyone else. Now they have camps for children whose parents run away, camps for people who kick and foam at the mouth, camps for people with big heads and people with little heads, camps for people with no visible means of support, camps for people chased off the land, camps for people they find living in storm-water drains, camps for street-girls, camps for people who can't add two and two, camps for people who can't add two and two, camps for people who live in the mountains blow up bridges in the night. Perhaps the truth is that it is enough to be out of the camps, out of all the camps at the same time. Perhaps that is enough achievement for the time being. How many people are there left who are neither locked up nor standing guard at the gate? (Coetzee 2015, 197–198)

Thus, the agency of a simpleton gets asserted in the end through the freedom that Michael K. experiences and which none of the other characters in the entire novel can lay claim to. He can escape because he manages to become invisible in the nondescript landscape that resists cartography. Hence, as a subject, Michael K. can neither be seen without the land he wants to become a part of or his ancestors, his mother and all the women who come before. He seems to become almost a part of the earth during several moments in the novel, for example, when children climb over him and he thinks he is one with the ground. He can tell the names of one bush from another by the smell of their leaves, he can smell the rainwater in the air. His knowledge (if it can be called so) hints at other possibilities of living. The book, fittingly, ends with a parable about drawing water from earth suggesting that perhaps there are minimal ways of establishing relationship with the land, with other forms of life. A similar relationship, although different in origins, configures Abani's narrative as well.

NATURE AS COMMUNITY IN CHRIS ABANI

In Chris Abani's novel *Song for Night*, the boy soldier My Luck forms a strange bond with his land and its water-bodies as he traverses through a nightmarish Nigeria during the last days of the Biafran war (1967–1970). However, the war described in the novel is not completely historical. Many critics have pointed out that it is more a fictional civil war, which has its roots in several civil wars, which occurred outside Nigeria, such as Sierra Leone, and Liberia. My Luck is fifteen years old and like numerous other children is trained to defuse landmines. And like all these other children, he is forcibly mutilated, his vocal cords are cut, so that he cannot scream if blown apart by a mine, thereby distracting the rest of the soldiers. My Luck, whose parents have been murdered, has to face intense violence throughout his short life as he is not only forced to bear witness to unspeakable events but also to take part in them. Nevertheless, as in all of Abani's works, the protagonist maintains a strong sense of ethics even when they are at the very limits of sanity and tolerance. In his journey across the inferno that is his homeland, My Luck desperately tries to find a philosophy or a set of codes based on which he can determine his actions and find the meaning of his existence. As many critics have pointed out, other than confronting difficult issues such as rape and wanton murder, Abani also deals with the horror of cannibalism during the Biafran war (Tunca 2013, 127–143). My Luck says, "I am mostly moving from one scene of past trauma to another, the distances between them, though vast, have collapsed to the span of a thought, and my platoon is ever elusive" (108). Thus, the translocation of the protagonist is not just across space and time inside the forest where he gets lost but also inside his own mind from one incident of his life to another. He faces his traumas one after the other and tries to give language to what has been happening to him and why. This is when his grandfather's Igbo stories become indispensable to him. Toward the beginning of his journey, My Luck looks at nature from the point of view of utility. Thus, rivers are the best way to keep close to habitation as well the fastest means of traveling, tree covers are necessary to hide inside, and so on. But toward the end, as he floats in a coffin on the same river seeking his fellow soldiers, the river is no longer a mere river. While being stuck in an island surrounded by the enemy, My Luck catches fish by remembering his grandfather's advice and then while eating the fish, he connects the river with Igbo life and as a result his own life. He thinks of the stories that concern the origin of the name of the river Cross. There were several tales regarding the name. For example, one story said that Igbos could be Hebrews who migrated to West Africa while carrying fragments of the Christ's Cross with them. Unsurprisingly, another tale speaks of the river spirit mami-wata's resistance to the building of a bridge by the Colonial Service Works Department. The engineer who tried eight times and failed is rumored to have muttered, "Bloody nigger river can't be crossed. I won't let it become my cross" (Abani 2007, 53). It is interesting how the river has to be attributed with the racial slur of "nigger" once it is seen as unmanageable, which was perhaps one of the points that My Luck's grandfather attempted to highlight. In order to feel connected with the river, he taught his grandson a song that he could sing over and over so that their voices mingled and My Luck no longer knew where his blood ended and the river began. This transcorporeality becomes even more significant when one realizes that the children who are being talked about are constantly under the threat of being torn into pieces by landmines that "civilized" Western countries would not use. Inevitably, they ended up in various parts of Africa because they are given away at extremely cheap prices in comparison to bullets and other arms. My Luck narrates how counting the dead after passing through mines is not simple since the shrapnel from mines tear bodies apart into several pieces. Assembling these pieces becomes a herculean task for the clean-up crew since many of the parts would refuse to add up in the form of the complete body of a human. Across this landscape, one could find groups of children lacking arms and legs and people with half faces but still holding on to some form of hope. These might also be the people who are later rejected by their own as zombies or ghosts. He says. "This is the enemy's cruelty—that much of the generation who survive this war will not be able to rebuild their communities. Even now it is not uncommon to run across groups of these half people holding onto life in distant parts of the forest" (Abani 2007, 37). When quotidian life consists of incidents such as these, the notion of water-bodies as repositories of the essence of different kinds of life including humans can create a much-required sense of belonging, a philosophical refuge. Since in that case, the body or the soul is not limited by what has been destroyed as a result of war. In his semi-autobiographical work, Abani discusses how even the face of a human being is not just his face in Igbo or Yoruba philosophy. Thus, the outer face leads to what lies inside a person but also through a dream-time to their ancestral lineage, to their culture, to the very soul of their land and people. Thus, even the concept of beauty is linked to an enmeshed and networked existence. Abani says:

> In Yoruba, Iwalewa is beauty and means the beauty of truth or even the beauty of existence. The word Iwa is best translated to mean existence, an eternal state,

being outside of time. Reality is held in IgbaIwa, the calabash of existence. Iwa is connected to an old idea that holds that immortality is the perfect existence, or better, a timelessness. It suggests that all temporality has ramification in an eternal cycle of existence—at an individual level, at a communal and lineage level, and in many ways at a planetary level. Everyone's Iwa is a part of the IgbaIwa, and the perfect balance between all I was depends on the singular balance of each. (Abani 2014, 16)

In a similar way, the story of the lake of fire and water is another notion that helps My Luck think through his trauma as he continues fighting to survive. The lake, which according to his grandfather is the oldest legend of the Igbo, is hidden in the fold of time. It is "the repository of human souls who are yet to gain access into the world" (54), guarded by fish and dolphins. It also lies in the center of the heart of Igbo people, thus linking the idea of community with a particular aquatic body. In a sense, it does not matter if at the end of the novel, My Luck is human or a ghost or something in between because he can go back to his mother and the lake. In that way, the lake is also an inner private space that My Luck inhabits. But it is a private space that he shares not just with his ancestors or other Igbo people but also various creatures. This space combats the complete transformation of My Luck into nothing more than bare life and memory becomes something that does not belong exclusively to humans. In an essay, "Creaturely Mimesis: Life after Necro Politics in Chris Abani's *Song for Night*," Sam Durrant reads Abani's novel through the lens of animism. Animism does not perform the simple task of mimicking the world in the sense that it represents it, animism in literature helps the reader understand that the world is connected and entangled with various forms of life other than human. A literary work thus becomes a way of identifying with the world and its beings and the novel becomes a transformative experience. As My Luck journeys through a world of magical realism, somewhere in between the dead and the living, his grandfather's philosophy of animism helps him stay connected with the world. Thus, Durrant comments: "Such an approach also moves us beyond the idea of art as an exclusively human activity and indeed toward an understanding of art as the renunciation of human sovereignty and the affirmation of our creaturely connections" (Durrant 2018, 185). Hence, the inherent anthropomorphism that is associated with a large section of literature is challenged and at the same time ecologically inflected indigenous belief systems are highlighted, bringing the entangled nature of postcolonial modernity to the forefront.

SHARED SUBJECTIVITY

In a sense both Michael K. and My Luck are trying to find a meaning in existence and the only way they can do it is through creating connections with their ancestors. For Michael K., the father's law does not help him escape; it is his mother's love for the land, which ultimately leads him to fulfill his desire of remaining outside camps. For My Luck, neither Christianity nor Islam and their philosophies can guide him through the inferno that he has descended into. In the end, only Igbo notions of the interconnectedness of the land, the water, the fish, and the human assists him in reaching a home where he can, as human or as spirit, find solace. The idea of subjectivity in both the cases is connected to nature. And perhaps as Mbembe has suggested, if humanity exists being in and through the world, it is perhaps possible to think of a relation with others based on the reciprocal recognition of our common vulnerability and finitude. This finitude is also the finitude of the human body in the way one sees it since in both the novels, it cannot be decided after a while where the body of the central characters ends and where the land or the lake or the river begins. The fish that My Luck consumes, he later thinks of as part of the human soul. Michael K. has a strange relationship with the vegetables that he grows, some of which he starts to think of as his children. Here, one can discuss the question of a new kind of subject, a subject that is not based on anthropocentric sovereignty. The human sovereign subject, as Tiffin and Huggan have pointed out, is almost always the privileged white male. The basis on which they assume this is that the western definition of what consists of the human is established on exclusion. The animal who cannot use language, or who does not possess language is what is "not-human." Again, the vast majority of people who are considered "barbaric" or "uncivilized" or "animal-like" are also as a result not-human. Hence, they say:

> European justification for invasion and colonization proceeded from this basis, understanding non-European lands and the people and animals that inhabited them as "spaces," "unused, underused or empty." The very ideology of colonization is thus one where anthropocentrism and Eurocentrism are inseparable, with the anthropocentrism underlying Eurocentrism being used to justify those forms of European colonialism that see "indigenous cultures as 'primitive,' less rational, and closer to children, animals and nature." (Tiffin and Huggan 2010, 5)

A large section of oppressed humans is thus systematically turned into "bare-life," thereby taking away all their rights and treating them simply as resources ready to be exploited. One way of theoretically countering this is also to displace the human itself from its central position and to place the human as one of the actants in a connected network of several creatures. This can also be done on the basis of shared suffering. The subject can thus be seen less a closed entity and more an opening where a gathering of beings can occur. Durrant in his essay had brilliantly suggested that even the biopolitical turn in postcolonial studies has failed to look into the spiritual dimensions of both the literature it sets out to read and the thinkers to which it returns but also to the creaturely dimensions of biopolitical thought. The creature is not only the sign of bare life, of a human being stripped of all cultural and spiritual pretentions, but also at least for dialectical thinkers such as Benjamin, the sign of a post-sovereign and post-human form of redemption. The creature marks the moment at which human alienation tips over into forms of trans-species relationality, into an awareness of our common creatureliness (Durrant 2018, 186). The affirmation of coexistence thus also affirms the instability of the subject. The subject can also be seen as a political formation, a myth or a story, that is carefully crafted. Just as the subject can be seen as an isolated human, limits can be drawn and broken to form communities and networks and there can be different, more ethical ways of looking at subjectivity. In both Coetzee and Abani's works discussed earlier, the story concerning the subject also concerns embodied existences and their various relationships. Michael's dependence on the land and his vegetables as well as My Luck's spirituality can be seen as political fables but then again, all subjects may be seen as such. For example, as Étienne Balibar has noted in *Citizen Subject: Foundations for Philosophical Anthropology*, the subject has to be seen through both subjection and subjectivation. Subjection is the willing obedience of the subject to an authority. On the other hand, subjectivation is related to the subject's responsibility and accountability. The human now makes the law, constitutes, and declares it. But, even then, since Balibar historicizes subjectivation by looking at the French revolution, the subject still remains a category based on exclusion. The essence of the subject remains man who is universal. One can note at this point, a certain statement that the medical officer makes when he tells Michael K., "The laws are made of iron, Michaels, I hope you are learning that. No matter how thin you make yourself, they will not relax. There is no hope left for universal souls" (Coetzee 2015, 166). This statement is ironic, since people such as Michael K. are excluded when one speaks of the citizen subject. But there might also be a suggestion that the subject can be thought of in other ways. It is not a closed category. Balibar also suggests that one cannot talk about the human subject without speaking of it as an ensemble of social relations and that can radically displace not just the way in which man has been understood but also the way essence has been understood. Thus, for Balibar, it is futile to question whether genus or essence precedes existence but what is necessary to think about human existence are the multiple and active relations, which individuals establish with each other through language, labor, love, reproduction, domination, conflict, and so on, and these are the relations that determine subjectivity. Therefore, he comes up with the notion of the transindividual to characterize the constitutive relation, which displaces the question of human essence while trying to talk about what exists between individuals by dint of their multiple interactions. The freedom of the subject may lie in these relations. Balibar, however, does not extend these relations to anything that is nonhuman. Both Coetzee and Abani base their notions of the subject on what is other than human. The use of the word "soul" in Coetzee is also interesting because unlike Abani, he does not base his novel on any philosophy of the soul. But the use of the phrase hints toward possibilities of reading the novel through that lens. In case of Abani's work, Major Essien nicknamed John Wayne (a reminder of America) subjects the children to his nonexistent survival manual that they eventually internalize. Their entire way of living their lives is controlled by these rules because of which every part of their existence becomes necropolitical. Ruptures emerge when My Luck is forced to commit crimes that he does not even comprehend, for example, when he is made to rape a woman on gun point. This is what makes him completely subjected to powers that he cannot question or negotiate with. The turning point comes when he turns against Wayne and his manual, thus changing the course of his life and resisting being turned into an object. He realizes that Major Essien is going to rape a seven-year-old child and he shoots at him thus killing both their commander and the girl, killing the one he claims was determined to turn them into animals. His trauma makes him inscribe crosses on his arms, creating his own personal cemetery. The first cross for when his grandfather died, then for his father and mother, his comrades, friends, lover, and the seven-year-old. The memories that he holds on to by inscribing them on his body also challenges the determinations made by authority regarding who is disposable and who is not. His body refuses to be inscribed by power in the way they want it to be used and discarded. He does to others what he perhaps wants to be done to himself. Similarly, when he finds a skeleton in a canoe, he tries to give it a burial. To quote from the book:

> I dig a shallow grave in the shifting sand, knowing it will be washed away in next year's flood. But that is unimportant. What is important is that this person be buried, be mourned. Be remembered. Even for a minute. Before I take the skeleton out of the canoe, I reach in and pull the cobweb gently free. I drape it over my own head like a cap and then lift the skeleton with ease, careful not to shake any bones loose. To come back complete, it is important that one leave complete. (Abani 2007, 58)

As he transitions from a human to a spirit, he unlearns the rules of the Commander, following the clues that his grandfather left him with. Both Michael K. and My Luck have to find their individual path with nothing to survive on but memories of association with land, with water, with nature, and with fellow humans. But in order to find their own identity, they have to form their own solidarities and choose a way of being in the world. The way the construction of their identity depends on these attachments pave ways for alternative possibilities of shared subjectivity.

BIBLIOGRAPHY

Abani, Chris. 2007. *Song for Night*. New York: Akashic Books.
———. 2014. *The Face: The Cartography of the Void*. New York: Restless Books.
Balibar, Étienne. 1995. *The Philosophy of Marx*. New York: Verso.
———. 2016. *Citizen-Subject: Foundations of Philosophical Anthropology*. New York: Fordham University Press.
Cadava, Eduardo, et al., editors. 1991. *Who Comes After the Subject?* New York: Routledge.
Coetzee, J. M. 1985. *Life and Times of Michael K*. New York: Penguin Random House.
———. 1986. *Foe*. New York: Viking Press.
———. 2015. *Waiting for the Barbarians*. New York: Random House.
Derrida, Jacques. 2006. *The Animal that Therefore I Am*. Translated by David Wills. New York: Fordham University Press.
Durrant, Sam. 2018. "Creaturely Mimesis: Life After Necropolitics in Chris Abani's *Song for Night*." *Research in African Literatures*, Vol. 49, No. 3 (Fall): 178–206. Academia.Edu.
Haraway, Donna. 2003. *The Companion Species Manifesto: Dogs, People and Significant Otherness*. Chicago: Prickly Paradigm Press.
———. 2008. *When Species Meet*. Minneapolis: University of Minnesota Press.
Huggan, Graham, and Helen Tiffin. 2010. *Postcolonial Ecocriticism Literature, Animals, Environment*. New York: Routledge.
Mbembe, Achille. 2019. *Necropolitics*. Durham: Duke University Press.
Poyner, Jane. 2009. *J. M. Coetzee and the Paradox of Postcolonial Authorship*. Burlington: Ashgate.
Tunca, Daria. 2013. ""We Die Only Once, and for Such a Long Time": Approaching Trauma through Translocation in Chris Abani's Song for Night." In *Postcolonial Translocations: Cultural Representation and Critical Spatial Thinking*, edited by Marga Munkelt, et al., 127–143. Amsterdam and New York: Rodopi. Academia. Edu.

Chapter 14

Colonialism, Capitalism, and Nature

A Study of Alex Haley's Roots *and Ngũgĩ wa Thiong'o's* Petals of Blood

Shivani Duggal

Colonialism is defined as a concept that involves "any sort of domination or assertion of control by one human group over another, often achieved by trickery and usually involving illegitimate means." The subjugation of one group creates an environment of psychological distress, discrimination, and dehumanization. The domination "extending political, economic, and sometimes even social power" leads to disparities and fissures within the society, which cannot be mended. Colonialists try to "[create] new iterations of their ways of life in distant settlements" (Page and Sonnenburg 2003, xxi) and attempt to coercively mold the colonized in accordance with their mentality.

As Sender and Smith assert, "profound social and economic changes . . . have involved brutality, disruption and suffering" (2013, 1) of humanity, due to colonialism. Colonialism involves not just brutally subjugating someone but also inculcates economic, cultural, and psychological domination of the subjugated. "Imperialism is itself a phase of capitalist development . . . [that] established political, economic, military, and cultural hegemony" (Rodney 2018, 13) over underdeveloped and subjugated people. The colonizers assert their domination and drastically transform the understanding of humanity and equality.

Max Weber defines capitalism as a "pursuit of profit, and forever *renewed* profit, by a means of continuous, rational, capitalistic enterprise" (1958, 17). Taking advantage of the present opportunities becomes the main focus, irrespective of the negative impact it would make on people. Individual growth and material gain become the primary concern where capitalism becomes "the *whole* system-social, economic, demographic, cultural, ideological"

(Mason 2015, 5). The capitalistic endeavors of people influence and engage innocent people into doing everything in accordance with elites, in order for them to sustain themselves in the society.

Capitalism creates a dichotomy between society according to their social and economic status. It leads to two different class formations that are "firstly, the capitalists or bourgeoisie who owned the factories and banks . . . and secondly, the workers or proletariat who worked in the factories of the said bourgeoisie" (Rodney 2018, 258). The distinct ways of living and earning livelihood create rift in the society. The economic hegemony involves the cultural and social hegemony too, where one group feels subservient to another dominant group. The feeling of being controlled and the constant surveillance affects the proletariats enormously.

Colonialism becomes another tool for capitalism, where capitalists economically take over another's economy and production, by colonizing them, in order to grow their trade. African continent was colonized by whites, who had capitalist motives behind colonizing Africans. Africa was an agricultural continent that sustained itself through farming, cultivating, harvesting, and trading the production. Whites took over their trade in the name of educating, civilizing, and teaching them better ways of leading their lives. The repercussions of inhumanely dominating the Africans are still reverberated and shown by their psychic disruption, identity crisis, and irresolvable complexities.

As Rodney elucidates, "capitalism and colonialism created the starvation, suffering and misery" and "chronic undernourishment, malnutrition and deterioration" (2018, 287) among Africans. The natural ways of living were totally disrupted as modernization, industrialization, and capitalism were introduced by colonizers in a pastoral space. Due to these self-centered industrial developments and growth in trade, colonizers robbed Africa of its strength, which was due to its community's solidarity and oneness. As Rodney further explicates, "Impact of colonialism (which) left so many villages deserted and starving, because the able-bodied males had gone off to labor elsewhere" (2018, 282).

Colonialism and capitalism affected the surrounding environment and nature too because colonialists exhausted the natural resources without restraint. In order to have excessive production for their economical growth, colonizers weren't concerned about the consequences that would prevail in Africa. The exploitation of nature deeply affected the Africans, since they were heartedly connected to it and also nature was the only source of their sustenance. The crops were cultivated according to colonizers' needs, farms were exhausted by excessive cultivation, and premature harvesting ruined the productivity of nature.

African continent has primarily been an agricultural land where human beings and natural environment were closely knitted for ages. Environment has always played two roles of being a protector and a destroyer, depending on people's treatment and response toward it. African's cultural repertoire has proven the role of nature as divine and a collective protector of humanity. Fulfilling all the basic needs and requirements of beings, nature has become a continuous source of livelihood and income for Africans. The homeliness felt by African community living amid nature was thwarted inhumanely with the intervention of capitalism.

Being holistically connected with nature, which was a sacred haven for Africans, they found solace in her warm hold. The unified feeling of oneness, collective understanding, and familial bond was celebrated within African community. The sanctity and purity of nature were revered by the inhabitants during cultural festivals and traditional ululations. The unified wholeness of Africa and its people was disrupted by colonizers who crossed that sacred periphery and claimed African land as their own.

The exploitation of nature and dehumanization of human beings are completely intertwined. Colonizers violated the protective boundaries of nature and then ill-treated Africans who were living peaceful life in that pastoral environment. Nature being a protective layer was forcefully deforested and thus it affected African inhabitants immensely. Due to colonizer's invasion, they were deprived of the basic necessities of life; which otherwise were provided by nature. Instead of tending their own farmlands, Africans had to work as wage laborers for colonizers so as to protect themselves from their wary.

Paucity of food and livelihood due to colonialism affected Africans' sense of identity, which was defined by its connection with nature. Africans used to "define themselves through the ways that they interact with nature" (Clayton and Opotow 2003, 9), thus the loss of nature as a whole led to their loss of self as viable human beings. The capitalist motives of colonizers were attained when they colonized Africa and exploited the agrarian resources and hard labor of Africans for their own economic benefits.

The aim of this chapter is to show how Alex Haley's *Roots: The Saga of an American Family* and Ngũgĩ wa Thiong'o's *Petals of Blood* incorporate the elements of colonialism, which is completely intertwined with capitalistic intentions. The chapter focuses on how colonialism affected natural environment due to colonizers exhausting the resources and deprived Africans by forcing them into hard labor, on their own lands. The capitalist motives of colonizers led them to take over African land, make colonies on Africans' sanctified farms, and trade their agricultural production for their economic benefits.

The chapter tries to illuminate how capitalism is intertwined and interconnected with colonialism and how both of these factors affected nature and human beings immensely. Capitalism became the primary motive of Europeans to colonize Africa, which had rich agricultural production that would increase colonizers' trade and economic position. Haley's *Roots* presents transportation of African natives to America through Atlantic, rampant slave trade, and years of exploitation of innocent Africans at the hands of slave owners. Ngũgĩ's *Petals of Blood* depicts the aftermath of colonialism and indecisiveness of Kenyans to utilize their independence for collective development, rather than individual pursuits.

The title of both the novels also puts light on nature's role being connected with the social and political surroundings. The "social and natural environment exert a roughly equivalent influence" (Clayton and Opotow 2003, 10) in the novels. The title *Roots* connotes the roots of a tree and roots of Africans' genealogy. Revisiting the ancestral past highlights the author's quest to trace the long-lost natural habitat of Africa, which was not yet encroached by predators. The title of Ngũgĩ's novel *Petals of Blood* exemplifies the marred past of Africans who had to sacrifice their labor, sweat, and blood for colonizer's profitable business. The petals marked red with blood can also annotate the sacrifices of Maoists to free their people from colonizers.

Roots: The Saga of an American Family is a novel written by Alex Haley in 1976, which is a genealogical investigation done by Haley himself in order to unfold his familial roots. The narrative is an amalgam of fact and fiction, which covers seven generations of Haley's family. Beginning the novel in an African village, Juffure, and ending the narrative in United States, Haley tries to portray the life of the protagonist Kunta Kinte's family in precolonial, colonial, and postcolonial times. The novel discusses slave trade, inhumane treatment of African slaves, labor exploitation, forceful sexual intercourse, plight of mixed race, and strife for space and individual identity.

Kunta Kinte, the protagonist, was born in the spring of 1750 and becomes the mouthpiece for the author to elucidate about the slavery system and the harsh implications it had on African's psyche. Spring season being the period of productivity, creativity, growth, and fervor; Kunta imbibed these character traits in his demeanor too. Kunta's abduction by African "slatees" for capitalist motive highlights on the loss of nature and mother earth at the hands of traitors and invaders. Haley shows the loss of agricultural productivity and compassionate humanity; and depicts the agriculture exploitation of nature and Africans in West for the sake of pecuniary benefits.

Kunta's childhood was spent amid natural environment, trees, forests, animals, herds, and cattle. Cattles were considered as most precious possession in African culture, which define a person's wealthy status in the African community. In order to appease nature, which was considered as divine, people even sacrificed their cattle during any natural calamity. The connection between nature and human beings in Africa has always been so strong but with the advent of imperialism, inhuman trials were perpetuated on human beings and nature simultaneously.

The portrayal of the arduous sea journey where the innocent Africans were being kidnapped from their native places and transported to a white's colony depicts exploitation of both nature and human beings. The spatial distance between Africa and West through vast Atlantic was violated via "Lord Ligonier," a British slave ship. Vast ships being the symbol of westernization and modernization annotate how modern capitalistic fervor of British disrupted nature's solace and shattered African community's composite identity. The migration from rural life to capitalized world due to colonialism led Africans to suffer oppression for long years.

The difficult sea journey was like "a nightmarish bedlam of shrieking, weeping, praying" (Haley 1991, 150) as hundreds of Africans were taken aboard and were deprived of the basic human necessities. They were chained together inhumanely, where rodents were constantly biting their naked bodies. One African who was being enslaved on the ship revolted and killed many whites but at last was decapitated and thrown into the sea. This reflects on the bravery and revolutionary spirit of the African in order to save his own people and land from being captured by predators and be disgraced.

The dehumanization of Africans in white man's land can be seen from the way Kunta Kinte's foot was cut off by his owners. The fragmented psyche of Kunta always reminded him of his tough times when he tried to free himself from the slave owner, John Waller. He was constantly tortured by the memory of his "blistered feet, the bursting lungs, the bleeding hands, the tearing thorns, the baying of the hounds, the snarling jaws, the gunshots, the sting of the lash, the falling ax" (Haley 1991, 335). The incident elucidates the barbaric and vicious treatment of any slave who tries to run from his or her slave owner's plantation.

The depiction of farmland, agricultural crops, and animals is in stark contrast, in accordance with the cultural and spatial difference. Africans celebrated harvest season, revered nature, and included nature in their festivities, whereas Westerners treated animals and crops they possessed inhumanely. They overused their farming lands for more production and trade, not entirely for consumption, like Africans. When Kunta Kinte started working as a slave in William Waller's house, he reminisced about his homeland and contrasted it with the ways of farming in West. He kept "Juffure's illusionary *secure inside* into its realistic *insecure outside*" (Osagie 2004, 397) so as to retain his sanity and strength. He felt disconnected from animals, nature, and agricultural work done at that dominating place.

Colonizers not only enslaved Africans but also deprived them of their true individual as well as communal identity. As is depicted in the novel, the slave owners gave new names to the slaves who worked under them and thus Kunta's name was changed to Toby. Loss of an African name that had cultural and familial connotation to it further annihilated the sense of self and led to psychic disruption. Kunta's name was taken from that of his grandfather Kairaba Kunta Kinte, who saved the village Gambia from draught by his belief and spiritual connection with nature. The connectivity Kunta felt with his name was thwarted that symbolizes his parting from nature and his familial community.

Tom Lea, a slave trader, bought Kizzy, Kunta Kinte's daughter, during a slave auction and raped her to beget a child for more labor force. The atrocities faced by African women have been portrayed by Haley that how African female slaves were priced according to their age and capacity to bore children. The crossing of the peripheries and forcing a woman to produce a labor force for selfish motives can be seen as how nature was exhausted and extensively used in white man's land for capitalistic development. It can be seen how exploitation of women signifies misuse of nature because nature is considered as mother figure and provider in Africa.

Tom Lea was a slave owner who was engaged in cock-fighting business and he trained his indentured slaves to fight for him and earned innumerable prize money. George, Lea's son born out of rape of Kizzy, also got entranced with the challenging sport and became chickens' trainer. He was popularized with the name Chicken George because of his amazing skills at training chickens, which implies the identity of an African as ephemeral and transient. Chicken George's identity was interrelated with that of the chickens, which he was tending and training. As soon as the importance of chickens for trade purpose vanished, his identity also withered.

In the last cock fight competition that Lea fought with an Englishman, Lea lost the most precious chicken, which fought in various fights and had brought enough wealth for Lea. Chicken's demise and Chicken George's migration to England away from his family delineate how Lea assumed him as useless without the task he was supposed to perform as a slave. Lea exploited chickens for his own monetary gain to attain a higher pedestal in capitalist society. He sacrificed birds, animals, chickens, and his own son Chicken George in order to own money and capital. Thus, it becomes clear that exploitation of man and nature goes hand in hand.

Tom Lea lost all his wealth in the last cock fight game and in order to pay off his debts, he sold his trustworthy slaves who worked for him for years. This comments on how "Niggers is the biggest thing they got" (Haley 1991, 259) and used them as possession which can be utilized according to the situational needs. Africans being considered as animalistic and barbaric by whites, "a mile long o' niggers bein' driv along in chains" (Haley 1991, 513) could be seen during slave trade and slave auction. Depriving them of unification with nature by constantly forcing them into unrealistic tasks like fighting in wars, slaves felt disconnected and lost in the new land, which was soon to become their home.

When Matilda announced that "Pres'dent Lincoln done signed 'Mancipation Proclamation dat set us *free*" (Haley 1991, 642), the newly found emancipation left the slaves perplexed. Kizzy's whole family stayed at the farm of their slave owner Mr. Murray and continued their farming tasks in order to sustain themselves. The shift from doing agricultural farm work for a slave owner to doing it for their own sustenance and growth, Africans familiarized with the work they did on an emotional level. Once again they felt connected with nature and tilled their own farms as they would do, if in Africa.

Africans "ascribe identity to natural forces and objects such as trees, animals, mountains, or winds, endowing them with intentionality, emotional response, subjective perspective, or simply spiritual significance" (Clayton and Opotow 2003, 8–9). Connectivity with nature once lost was presumed again and slaves started identifying themselves as viable beings in an American society. The shift from having an African ancestry toward an Afro-American identity symbolizes cultural, traditional, and spatial shift of slaves. Newly freed slaves couldn't just do farming and harvesting; they had to adapt and accommodate themselves in the new working environment.

Identity crisis was faced by slaves who were born in America itself on the plantation of their slave owners. They found it difficult to connect with either their African culture, which they didn't interact with at first hand, or the American urbanization, which was not taking them into consideration. Lack of belongingness prevailed for Afro-American slaves but as soon as they were freed from years of bondage, they embarked on a journey to make space for themselves in this new land, where they then belonged. The identity transformation from native African to Afro-American asserts "development of a distinctive African-American culture" (Meritt 1977, 212).

The struggle ensued for the newly realized Afro-American identities but the plight of Afro-Americans improved because of their perseverance and continuous hard work. Kunta Kinte becomes a symbol of "richly textured fabric of identities-in-the-making" (Osagie 2004, 395) that explicates about innumerable slaves who carved a space for themselves in their colonizer's new world America. The novel inculcates the "string of dispersed realities and fractured memories that have produced African American identities within the *othered* space of the New World" (Osagie 2004, 394).

Haley went on to explore about his ancestral, traditional, cultural, communal, and familial "roots" in order to connect with the essence of his identity. He was encouraged to do so because of the narrative of Kunta Kinte's struggles being passed on to seven generations, so that the true African identity never withered from their mind and blood. Haley felt impacted by this repetitive narrative of Kunta and embarked on a journey to find his roots and from where did his identity originate. The narrative about exploration of ancestral "roots" "helped people understand the importance of their individual past"

(Meeks 1992, 21) in order to resuscitate their lost world, nature, memory, and compositeness.

The oral narration of Kunta's journey by a "griot" from being a free African to being an indentured labor highlights the traditional ways of narrating stories in African culture. "A griot is a member of a hereditary caste of praise singers, poets, genealogists, storytellers, musicians, and oral historians . . . . Griots have been known to memorize the entire genealogy or family history of everyone in an entire village going back for centuries" (Alexander and Rucker 2010, 47–48). They are revered in African culture for being the repository of their rich cultural past. The narrator went to Juffure village and with the help of a "griot" who were "walking archives of oral history" (Haley 1991, 674) kept the "roots" intact and proved the authenticity of the narrative.

When Alex Haley reached Juffure village where Kunta Kinte had spent his sixteen years, before he was abducted, Haley felt plethora of emotions rush through his body when he became a part of an oldest ceremony named "The laying on of hands." The traditional ceremony where the native women gave their babies for the narrator to hold one by one and said, "Through this flesh, which is us, we are you, and you are us!" (Haley 1991, 680). The narrator felt emotionally connected with his nativity, which was untarnished by modernity, urbanization, and capitalism. The connection with nature and her pure beings led Haley felt mesmerized and enlightened.

Petals of Blood by Ngũgĩ *wa Thiong'o*, a prominent Kenyan writer, is situated in the postcolonial times when Kenya gained independence from white colonizers. As colonizers have introduced urbanization and modernization in African land, its influence was still felt in the postcolonial Kenya. People have become materialistic, and rather than focusing on collective growth, they moved toward individual development. Ngũgĩ portrays "neocolonialism as a form of prostitution of the nation by elites" (Mwangi 2004, 67), which comments on the dichotomy between proletariats and elites. The oneness of Africans has been lost and the inhabitants are hanging in between the two opposite mental spaces, one of being traditional African and another of being modernized African.

The four main characters in the novel are Munira, Wanja, Karega, and Abdulla; and the narrative revolves around their individual journey toward groping with changed reality. The novel is set in a rural village of Ilmorog, which was still preserved and not contaminated by the urbanization. Ngũgĩ writes the novel keeping Mau Mau Rebellion as its background, when Maoists fought against colonizers in order to protect their land and nature. The difference between pastoral Ilmorog and modernized Nairobi, where elite Kenyans lived, comments on the few protected spaces in Kenya, which were yet to be dominated by capitalism and modernization.

Munira was a son of a rich landowner, Ezekieli, who had many farm laborers tilling his farms. Despite having family's stable wealth and lands, Munira felt unsatisfied and perplexed about his identity in Limuru. He came to Ilmorog as a teacher but with the primary motive of finding recluse in a solitary rural space, away from the humdrums of urbanization and modernization. While trying to educate children of Ilmorog, Munira was unable to become part of the life and people of Ilmorog because of his preconceived notions. Although he tried to connect with the pastoral space but his mind always dwindled between village and city.

He, being an outsider who has witnessed capitalistic growth and modern transformation of Kenya, couldn't grasp the significance of "petals of blood." When children questioned him about them, Munira plucked it and "now it was just a red flower" (Ngũgĩ 2005, 26). It symbolizes that education, which Munira was giving to the students, was detaching them from their original roots. Many "youth [were] running away from the land" (Ngũgĩ 2005, 23) to the capitalist world to strive for earning wages from menial labor. The "blood" that was in their veins was sucked out by elite Kenyans and led them dive deep into the vain darkness of poverty; thus transforming them into soulless beings.

Munira always strived to be a part of Ilmorog community but couldn't because in his heart he never questioned or considered whites as devastators. Munira became jealous of Karega, his teaching assistant, because of his enthusiasm for being concerned about Ilmorog's development and helping villagers. Karega's Maoist brother killed many capitalistic and authoritative politicians as a reaction against their selfish acts. Munira questioned his identity in Ilmorog and realized his unwillingness to be one with pastoral life of the inhabitants, who were only interested in agricultural work and not educational success.

Karega was in love with Munira's sister, Mukami, but later she committed suicide because of the domination of her father. Since Karega's brother was a Maoist, Ezekieli dissuaded her from continuing the relationship with Karega. Mukami was remonstrated for working in farms with other laborers and indulging in menial works. She could not grasp the modernized and urbanized Kenya and died in order to resist conforming to the new ways of living. It emphasizes that how people were discouraged to connect with nature and were forced to modernize in order to avoid subservience.

Karega taught the village children to be aware about the reality and not just remain engrossed in education for vain. Munira considered his "educational reforms too radical and urges Karega to be more conservative" (Podis and Saaka 1991, 114), which symbolizes the intellectual difference between both of them. Having been through different experiences, Karega was aware about the guerilla war and its significance for unprivileged natives. Eventually by the end of the novel, Karega's rebellious nature led him to become a "politically conscious activist and trade unionist" (Mwangi 2004, 67).

In his arduous attempt to bring equality and improvement in the lives of villagers, he traveled together to Nairobi along with them as a spokesperson of Ilmorog's unpleasant situation at the hands of corrupt politicians. Independence from colonialism introduced two different classes that are proletariats and bourgeoisie in the African community, which was so closely knitted in precolonial times. Some Kenyans became collaborators with colonizers to subjugate their own people for capitalistic motives. This created rift between Kenyans who were exploiters and Kenyans who had their compositeness and natural environment intact.

When industrialization and capitalism made its way through Ilmorog, it affected the agriculture production drastically. The capitalists mortgaged native's pious agricultural lands and divided them with fences. They claimed the lands and asked the villagers to pay loans to free and reclaim their lands. This demonstrates that how capitalism not only deprived people of basic amenities but also exacerbated the natural environment. Nature was dominated, exploited, and divided by capitalism and colonialism as were the people inhabiting it. Land's division illustrates the division of Kenyans done on the basis of their class and economic status.

Wanja wanted to protect her grandmother's land from being occupied by capitalist Kenyans and started a brothel there. She saw the dominating capitalist Kenyans usurping innocents' lands and she was forced to sell her body in order to protect her ancestral land. Like the land being divided in "several Ilmorogs" (Ngũgĩ 2005, 333), Wanja's body was also divided for the capitalist males to overpower. The connection between Wanja's body and Ilmorog's land highlights the exploitation of both human beings and environment due to capitalism.

Ngũgĩ portrayed Wanja as connected with nature; who binds Ilmorog together as a community. She impacted the lives of Ilmorog's inhabitants in optimistic manner. Her arrival in Ilmorog brought fervor, happiness, and productivity all over the village. She assisted Abdulla in growing his business, helped Joseph (Abdulla's helper) to pursue education, celebrated in all festivities, aided her grandmother in chores, and enhanced the emotional intellect of Munira. She becomes a unifying symbol in the novel, like nature unifies Kenyans while harvesting festivities and cultural merriment.

Wanja being the protector of her unified community, sacrifices her sanctity to keep her ancestral land sanctified. As Munira states, "She is that bird periodically born out of the ashes and dust." This shows that how Wanja did not remain in the peripheries of male domination and rather she "owns houses . . . fleet of matatus . . . fleet of big transport lorries" (Ngũgĩ 2005, 334). In contrast to her lay factory laborers who were groping with "new owners, master-servants of bank power, money and cunning" (Ngũgĩ 2005, 332) politicians, who were establishing new factories and enhancing tourism in New Ilmorog for gaining more money.

Wanja's grandmother, Nyakinyua, was revered in Ilmorog because of her determination and enthusiasm for protecting her people. She constantly prayed for rain to come so that the agricultural crops are harvested with full celebrations. In African culture, nature is considered to have divine power, which acts in accordance with people's deeds. If the community is solidified and interconnected to each other harmoniously, nature shows her acceptance by giving timely rains for good agricultural production. Since Kenyans were divided for monetary endeavors, villagers had tough time dealing with the difficulties surrounding natural changes.

Nyankinyua celebrated rains by making intoxicating drink from a Thang'eta plant that grew in Ilmorog. The drink was enjoyed by inhabitants and Wanja decided to sell the traditional drink at Abdulla's shop, a member of Mau Mau Rebellion. When influential businessmen understood people's attraction toward the Thang'eta drink, they claimed it and built up a factory for its production. The "capitalist appropriation of the drink," which was nature's gift and marked Ilmorog's celebrations, was transformed into a business product that comments on loss of celebration and happiness in Ilmorog. Karega being the rebellious spirit struck against it and tried to "reestablish Theng'eta as a force to be controlled by the people" (Podis and Saaka 1991, 67).

Industrialization in Ilmorog destroyed the farming lands and forced many farmers into working as factory laborers. The menial job with menial income affected the way Ilmorog worked, as earlier the community was agriculturally based but now they were working only as "hands." The manipulation of farmers exemplifies the exploitation of nature at the hands of factory owners for capital growth. Karega also worked in the factory only to realize the ill-treatment of laborers and misuse of power by Kenyan businessmen. Due to urbanization, road and factory construction, modernization, and capitalism, Ilmorog lost its essence even after being freed from colonialism.

Wanja's brothel can be contrasted with Theng'eta Breweries that Kenyan elites started in Ilmorog. Wanja resisted any domination from the male elites who were the leading figures in postcolonial times. She worked as a prostitute, which can be seen as a resistance from her side against male chauvinistic and capitalistic society. New Ilmorog that had roads, factories, business centers, and bourgeoning tourism followed only one law, which was "You eat somebody or you are eaten" (Ngũgĩ 2005, 345). Wanja was free to work voluntarily, whereas the paid wagers working in the factories were dependent on businessmen.

Abdulla was an active member in guerilla war but later migrated to pastoral Ilmorog in order to start his life anew. "Mau Mau was a militant response to years of frustration at the refusal of the colonial government to listen to demands for constitutional and land reform" (Maughan-Brown 1981, 3–4). He lost his leg during the rebellion, which shows the sacrifices proletariat Kenyans did for improving their meaningless lives. Abdulla came with a donkey that later became the only target of a plane crash in Ilmorog and died on the spot. Abulla's sole connection with donkey shattered due to modernity, which made its way in Ilmorog. Death of the donkey comments on the brutal treatment of animals and nature simultaneously due to budding capitalistic fervor in Kenyans.

Members of Parliament and political representatives of Ilmorog, who were responsible for inhabitants' well-being and development, abused their power for their self-development. Nderiwa Riera was one such politician who earlier spoke for "nationalization of the major industries and commercial enterprises; abolition of illiteracy and unemployment and the East African Federation as a step to Pan-African Unity" (Ngũgĩ 2005, 208) but later became a demagogue. Being part of Kenya, he exploited his own people, capitalized his own lands, and materialized the bounties of nature.

The businessmen established different organizations in New Ilmorog, which further divided the community on different levels. The organizations became "the most feared instrument of selective but coercive terror in the land" (Ngũgĩ 2005, 222). The coercion faced by the farmers at the hands of the capitalists shows how they were terrorized by powerful and influential politicians. The land and farms were sacrificed and transformed into roads, which became the symbol of trade and business growth.

Occupying the lands of inhabitants and claiming it as their own, the fight for land becomes the central discussion in the novel. As Ngũgĩ writes, "Land is presented as salvation, as a soul, as a woman, as God, the subject of prophecy, the basis of cultural and political identity" (2005, xiii). Wanja becomes a symbol of postcolonial Kenya, which has to overcome all destructions and grow steadily. The cultural significance of land as being the binder of communal identity transformed into the divider of Ilmorog, through construction of roads.

The "rhythmic balanced proportion of rain and sunshine" helped the people of Ilmorog to determine the expected production. Nature played a vital role in their lives because the village was dominantly agricultural. The "sporadic rains or by a continuous heavy downpour" (Ngũgĩ 2005, 38) affected the crop production and inhabitants always remained anxious about the nature's surprise. After Kenyan independence, the "virgin common land or unclaimed grassfields" (Ngũgĩ 2005, 11) turned into a wasteland, which could be seen as nature's revenge against her exploiters.

The title of the novel symbolizes Wanja's virginity, which had to be sacrificed in order to reform Ilmorog's situation. She is captured and raped by

Kimeria, a businessman, when she along with villagers traveled to Nairobi to demand basic human rights. As a woman's virginity is attached metaphorically to "blood," Wanja's sanctity was lost as she further became a prostitute. Even when Wanja left Ilmorog for a short span of time, nature drastically deteriorated. Throughout the novel, Wanja is compared with nature and her loss of virginity can be seen as the loss of sanctity of the agricultural land and nature due to imperialism and capitalism.

Africans' connection to "nonhuman natural environment, based on history, emotional attachment" (Rodney 2018, 10) explicates how they engage with their surroundings and nature with respect. Colonialism affected that collective conscious of Africans and disrupted the emotional connect Africans felt with nature. It rather divided people in accordance with their capital income; which are peasant class, politicians, and capitalists. The division of nature by separating once unified land and natural resources was possible due to "agents of repression from among the colonial victims themselves" (Rodney 2018, 172) like "traitees" and African collaborators. It further divided the African community on the basis of their involvement in the colonialism and exploitation of their own people.

In *Roots*, there were many collaborators who engaged with whites, kidnapped their community's people and sold them for monetary gains. Similarly, in *Petals of Blood*, many Kenyans collaborated with whites and thus gained power over other rural Kenyans after colonialism ended. This comments on how community gets divided because of its own people, who become attracted with the idea of authoritative power and monetary incentives. The land too was divided for construction of roads, railroads, working places, and camps of whites, who slowly entered into the divine space of the village and stole its crops, youth, and unity.

In *Petals of Blood*, inhabitants like Njamba Nene, Nyakinyua's husband, who fought for the rights of Ilmorog and its people, can be contrasted with Munoru, who was corrupt and tried to devastate Ilmorog's people for selfish motives. Whereas, the different ideology of Africans can be seen in *Roots*, where people like Kunta Kinte can be contrasted with the "traitees" who kidnapped him instead of fighting against whites. The domination of whites during colonialism and further misuse of power by authoritative Africans after colonialism comments on the division of communal oneness and disparities between two contrasting ideologies.

Thus, it can be concluded from the given analysis of two novels that Alex Haley described colonialism and capitalism in the colonial and postcolonial times, whereas Ngũgĩ *wa Thiong'o* discussed the aftermath of colonialism and how Kenyans coped up with the changing world. Both the writers critically analyze and portray the authoritative and capitalistic people including both whites and Africans. The exploitation of nature and Africans continues amid colonialism, capitalism, modernization, industrialization, and urbanization. Contrasting ideologies delineate the fissures in society, which can never be mended, restored, and reconstructed.

BIBLIOGRAPHY

Alexander, Leslie M., and Walter C. Rucker, Jr. 2010. *Encyclopedia of African American History* [3 volumes]. California: ABC-CLIO.

Clayton, Susan, and Susan Opotow. 2003. *Identity and the Natural Environment: The Psychological Significance of Nature*. London: MIT Press.

Haley, Alex. 1991. *Roots: The Saga of an American Family*. London: Vintage.

Mason, Paul. 2015. *Post Capitalism: A Guide to Our Future*. London: Penguin UK.

Maughan-Brown, David. 1981. ""Mau Mau" and Violence in Ngugi's Novels." *English in Africa* 8, no. 2 (September): 1–22. Accessed August 16, 2020. www.jstor.org/stable/40399033.

Meeks, Rob. 1992. "Alex Haley: Remembered." *History News* 47, no. 3 (May/June): 21. Accessed August 16, 2020. www.jstor.org/stable/42655754.

Meritt, Carole. 1977. "Looking at Afro-American Roots." *Review of Roots*, by Alex Haley. *Phylon (1960–)* 38, no. 2 (2nd Qtr): 211–212. Accessed August 16, 2020. doi: 10.2307/274684.

Mwangi, Evan. 2004. "The Gendered Politics of Untranslated Language and Aporia in Ngugi Wa Thiong'o's "Petals of Blood"." *Research in African Literatures* 35, no. 4 (Winter): 66–74. Accessed August 16, 2020. www.jstor.org/stable/3821204.

Ngũgĩ wa, Thiong'o. 2005. *Petals of Blood*. London: Penguin Books.

Osagie, Iyunolu. 2004. "Routed Passages: Narrative Memory and Identity in Alex Haley's "Roots"." *CLA Journal* 47, no. 4 (June): 391–408. Accessed August 16, 2020. www.jstor.org/stable/44325232.

Page, Melvin Eugene, and Penny M. Sonnenburg. 2003. *Colonialism: An International, Social, Cultural, and Political Encyclopedia*, Volume 1. California: ABC-CLIO.

Podis, Leonard A., and Yakubu Saaka. 1991. "Anthills of the Savannah and Petals of Blood: The Creation of a Usable Past." *Journal of Black Studies* 22, no. 1 (September): 104–122. Accessed August 16, 2020. www.jstor.org/stable/2784499.

Rodney, Walter. 2018. *How Europe Underdeveloped Africa*. New York: Verso Books.

Sender, John, and Sheila Smith. 2013. *The Development of Capitalism in Africa*. UK: Routledge.

Weber, Max. 1958. *The Protestant Ethic and the Spirit of Capitalism*. Translated by Talcott Parsons. New York: Scribner's.

Chapter 15

Beyond the Dichotomy of Humans and Animals

Situating Ecology in Coetzee's Writings

Bipasha Mandal

If we are to trace the emergence of different branches of biology, we would find that ecology is a relatively new branch of that science, and in its infancy, it was concerned with conservation. Conservation has always been a concept that had originated in the northern countries, especially Europe, and the agenda that it followed (and still does) is deeply linked with colonialism. It goes without saying that in recent light, in the sense that now that Western hegemonic ideologies are being questioned not only from the center but also from the margins, the environment-related questions of different regions are being reevaluated and reexamined. The academic discourses around the environment question concerning South Africa or other ex-colonies aim at moving beyond the Western colonial political thought and influences that inevitably shaped the ecological arena of the colonies. The change of political power in the ex-colonies also tries to steer the environment question in a direction that best suits its propaganda. In South Africa, the country's transition to democratic government also initiated the rethinking of environmentalism in a way that it became people-centric, and subsequently moving away from the conservation centric environmentalism, which had colonial roots. The people-centered environmentalism in South Africa as we know it today had also much to do with an activist organization called Earthlife Africa, initially conceived of as a South African version of Greenpeace. D. A. McDonald's *Environmental Justice in South Africa* opens with one of the most direct attacks at colonialism, and indirectly at anthropocentrism. He argues that "flora and fauna were often considered more important than the majority of the country's population" (McDonald 2002, 1). Broadly speaking, any discussion on postcolonialism is bound to be anthropocentric; however, the people-centric environmentalism that emerged in democratic South Africa fails to shatter or even partially crack the colonial ideals. He further argues that the environmental justice literature of postcolonial Africa fails to be homogenous, and even though it strives to be people-centric, it fails to move beyond the ideological constraints that are so deeply colonial, an ideology, from its very conception, had been built on the distinctions of race, class, and gender. It would be naïve on our part if we are to only accept only one dimension of such ecocriticism. Anthropocentric ecocriticism will inevitably create friction between the tendency to value human needs and to accept the environment as is with its own sets of values. If the differences are approached within a democratic framework of a country it would open up dialogue, and this, in turn, can create ecocriticism that is distinctly South African.

In this chapter, I read two fictional narratives by the South African writer, J. M. Coetzee, *The Lives of Animals* and *Disgrace*, both of which weave the question of ecology into the narrative plot, and offer a postcolonial perspective that is ambivalent in its stance against the northern power. Shohat has pointed out that Coetzee's writings designate "critical discourses which thematize issues emerging from colonial relations and their aftermath, covering a long historical span (including the present)" (1992, 101). I would like to explore, and answer to the best of my abilities, some of the ethical questions raised in both the texts regarding animal rights, the relation of human with nonhuman animal, where the line that makes the distinction between man and animal becomes blurred, and the possibility of transcendence beyond the dichotomy of the human and nonhuman animal. Furthermore, I would also like to point out the limit of the values assigned to ecology through fiction. Coetzee's narrative not only incites the readers to resist cruelty to animals on a thematic level, on the other hand, it also invites the reader to reconsider the terms and ideologies that this resistance is based on. This dual-mode of resistance is what Jonathan Bate has termed as "ecological work" (2000, 200). Bate has argued that any kind of eco-text has the potentiality to explore the relationship between the human and the nonhuman environment through what he has called an act of mimesis, of getting close to nature. The philosophical exploration of the questions regarding animals raised in Coetzee's narratives succeeds in decentering what has been rightly termed by Jonathan Skinner as the "habitual configuration" (2003, 5). He further argues that imagining endangered species is a useful act of language and that writing about it points out who is endangered, decentering the habitual configuration, and it is precisely this decentering, which could be used as an exemplary example of an ecological goal that an aesthetic discourse achieves. Ecocritics pose a challenge to modernity's thinking of the shared relationship between the humans and the nonhuman environment, and postcolonial critics in particular call for the rethinking of methodologies used for the environmental issues in the ex-colonies, especially since the ecological relation to a modernity is built on and rooted in colonialism. Benita Parry has commented that Coetzee's novels "interrogate colonialism's discursive power" (1999, 150) and this power tries to reinvent the ecological framework in South Africa.

The Lives of Animals (1999), which is a textual hybrid, part fiction part academic lecture, a meta-narrative, was a two-part lecture delivered by the writer himself as a part of the 1998 Tanner lectures at Princeton University, thereby we can safely assume the fictional character of Elisabeth Costello to be an alter-ego of Coetzee himself. Costello's lectures focus on an important ethical issue: the way humans treat animals. Costello's academic colleagues and fellow human beings commit "crime of stupefying proportion" (Coetzee, 1999, 69) in their treatment of animals. Surprisingly, Coetzee makes implicit in the text a subtle critique of the power structure. Costello's arguments about animal rights may seem cogent in the overt statement of extending our sympathies to the animals; however, the covert meaning of the statement, that we should not eat meat, may seem like it is coming from a professor who is dissociated from the power and class structure that ultimately shaped the world. Norma, Costello's daughter-in-law, rightly remarks, "a dietary ban is a quick, simple way for an elite group to define itself. Other people's table habits are unclean, we can't eat or drink with them" (Coetzee, 1999, 49). Costello also fails to take into account the cultural differences that prevail and something that shapes the people's outlook toward animals and their rights. O'Hearne rightly raises the question of "claiming universality for what are simply its own standards" (Coetzee, 1999, 60). By this claim, I do not mean that obligation to animals is a Western ideal, on the contrary, "the first travelers to South America encountered settlements where human beings and animals lived higgledy-piggledy together" (Coetzee, 1999, 61). Donald Worster in *Reinventing Nature? Responses to Postmodernist Deconstruction* (1995) has argued that nature works and sustains itself by the principle of interdependency, and this is something that was not invented in modern times. He also goes on to argue that the earliest cultures also discovered the interdependency of all living creatures. What I intend to highlight here is that Costello by claiming that the West has access to an ethical universal to which other cultures are blind, or the fact that the other cultures do not adhere to what the West believes to be the standard ethical boundaries, gives the West a free rein to establish a self-proclaimed superiority. While being sympathetic to Costello's noble cause, as McDonald has argued, resistance is what they will inevitably face if the fad from this particular perspective is imposed on other cultures.

Even though no direct connection is made in Costello's arguments to South Africa, the solemnity with which she makes the generalizations in the novel, and the seeming universality of her claims, relates her commentary to the African ex-colonies, by means of association. While talking about the relation between humans and animals, and what sets humans apart from the nonhumans is the Enlightenment-bred reason. The sheer extent of our intellectual prowess is greater than any other species on this planet, and this is exactly what gives humans the required validation that makes humans superior; the intellectual capacity of humans makes us capable of comprehending the concepts of life and death. However, by claiming reason to be the deciding agent here, she becomes ignorant to the racialized other, in the sense that segregation and discrimination among the humans had been enforced based on the same principle. Discrimination against black bodies is just another weapon of the empire for subjugating the people from the colonies. Not to mention, the power hierarchy also discriminates on the basis of gender, class, and even caste. The ecological framework that the West drew and tried to impose its own principles on the rest of the world was mainly focused on conservation. The racialized others were never admitted into the category of the man so that they could be categorized as the undesirable animals, while animals and plants could be marked off as desirable and hence worthy of conservation. McDonald rightly remarks that Coetzee's narrative brings out "the inconsistency that permits the dominant society to mark off nature preserves and device ecological programs in the world outside those preserves, while at the same time condoning and enhancing systems of industrial meat production, animal experimentations, etc." (2002, 9). Costello undertakes a radical critique of what Wolfe (2003) has called "the discourse of species" that has dominated Western intellectual thought for centuries. The discourse of species, also known as speciesism, uses the argument that one species claims superiority over other species by using the other species as foils; from an etymological perspective, this argument also highlights how certain humans claim superiority over other humans, for example, the relation between the words "*chattel*," "*cattle*," and "*capital*." Wolfe has rightly remarked, in "[the] humanist discourse of species [that] will always be available for use by some humans against the social other or *whatever* species- or gender, or race, or class, or sexual difference" (2003, 8). Costello tries to the best of her abilities to expose the ugliness of modern people as ethically bankrupt people. In the second part of the book "The Poets and the Animals" Costello attacks the philosopher's language, which is immersed in what Adorno and Horkheimer (1984) called "the dialectic of enlightenment":

> I am not sure I want to concede that I share reason with my opponent. Not
>
> when reason is what underpins the whole long philosophical tradition to which he belongs, stretching back to Descartes and beyond Descartes through Aquinas

and Augustine to the Stoics and Aristotle. If the last common ground that I have with him is reason, and if the reason is what sets me apart from the veal calf, then thank you but no thank you, I'll talk to someone else. (Coetzee, 1999, 66–67)

The killing of animals for the food industry and the blindness of the majority are controversially compared directly to the killing of Jews and the blindness of the German population under Hitler; however, through implication, the lectures also connect to both of these the history of colonial conquest and exploitation. In her comments on *Gulliver's Travels*, Costello alludes to the inevitable violence that was inflicted on the local inhabitants after the European explorers landed on the foreigners' lands and took control there, and she concludes by pointing how animals and humans have been the objects of slaughter for every civilization that was conquered, and the humans were met with the same fate as the animals, either slaughtered or enslaved. However, the metropolitan audience to whom Costello delivers the lecture seems to completely forgo the implicit colonial implication. The audience asks questions and engages in debates by asking reasonable and rational questions, without acknowledging the colonial past of their own country where native Indians were slaughtered, and Africans used as chattel. McDonald has argued that the audience exhibits the "characteristic of an organized, rational world and illuminating its values" (2002, 8). Through Costello, the writer questions the ethical grounds of American dominance, and the willful ignorance of its people regarding the subjugation of the "human-other," a mechanism that is deeply rooted in the European empire, which also sustains the current power structure and maintains the status quo.

In *The Lives of Animals*, Costello tries to bring to the foreground what she terms as our "sympathetic imagination" (1999, 35). We can say with surety that we certainly do not lack such imagination; works of fiction make that pathway easier for us to tread on. Likewise, she argues, "If I can think my way into the existence of a being who has never existed, then I can think my way into the existence of a bat or a chimpanzee or an oyster, any being with whom I share the substrate of life" (Coetzee, 1999, 35). This "sympathetic imagination" finds its best expression in two poems, Rainer Marina Rilke's "The Panther," and Ted Hughes's "The Jaguar." She argues that in "Rilke's poem the panther is there as a stand-in for something else" (Coetzee, 1999, 50), and goes on to consider that Hughes's poem is the superior of the two because "in these poems we know the jaguar not from the way he seems but from the way he moves. The body is as the body moves, or as the currents of life move within it. The poems ask us to imagine our way into that way of moving, to inhabit that body" (Coetzee, 1999, 51). Her chosen preference for the poem "The Jaguar" over Rilke's perhaps lies on the fact that this poem resonates with her own political conviction, which is almost the same as the jaguar's wildness, reinforcing yet again the anthropomorphic quality of Hughes's poem. The union between humans and animals that Costello so eloquently has expressed could be achieved through the sympathetic imagination; however, the obvious inquiry rising out of this argument would be that the union that she talks about is a matter of representation: "Isn't that what is so suspect about the whole animal rights business: that it has to ride on the back of pensive gorillas and sexy jaguars and huggable pandas because the real objects of concern, chickens and pigs, to say nothing of white rats or prawns, are not newsworthy?" (Coetzee, 1999, 55). Her son's remark ironically reminds us that her lecture with all its motivating persuasion remains just a work of fiction.

Costello reads Kafka's "A Report to an Academy" and during her comments she implies that Kafka drew an analogy between racial discrimination and the speciesism to which Red Peter, the ape in Kafka's story is a victim: "If Red Peter took it upon himself to make the arduous descent from the silence of the beasts to the gabble of reason in the spirit of scapegoat, the chosen one, then his amanuensis was a scapegoat from birth, with a presentiment, a *Vorgefühl*, for the massacre of the chosen people that was to take place soon after his death" (Coetzee, 1999, 26). What Costello fails to take into account from Kafka's story is that Red Peter, the ape, even though fully conversant with European ideals and rationalism, occupies a marginal position by the very figure of his body. Red Peter has been described as the specter of the Jews in European society, but the fact that Red Peter comes from "the Gold Coast," a place not in Europe, makes him a metaphor in many ways for the colonial subject who "must confront the ironies of his acculturation in European values" (Barney 2004, 4). Caught between the two dichotomies, Red Peter chose a third option, his "way out" which Deleuze and Guattari have described as "becoming animal," which could be described as the erasure of traditionally defined human features, making way for a mode of existence that could surpass the binaries that shape the human. As Costello remarks, "Kafka saw himself and Red Peter as hybrids, as monstrous thinking devices mounted inexplicably on suffering animal bodies" (Coetzee, 1999, 30). For Coetzee too, the "way out" remains a third alternative, albeit vague in nature, perhaps also a site for hybridity with tangible benefits.

Her talk takes a surprising turn when she makes the atrocious analogy between the slaughterhouses and the Nazi death camps, and by rules of implication, between animals and the Jews. She contends:

> The particular horror of the camps, the horror that convinces us what went on there was a crime against humanity, is not that despite a humanity shared with their victims, the killers treated them like lice. That is too abstract. The horror

is that the killers refused to think themselves into the peace of their victims, as did everyone else. They said, "It is they in those cattle-cars rattling past." They did not say, "How would it be if it were I in that cattle-car?" They did not say. "It is I who am in that cattle-car." They said, "It must be the dead who are being burnt today, making the air stink and falling in ash on my cabbages." They did not say, "How would it be if I were burning?" They did not say, "I am burning, I am falling in ash." (Coetzee, 1999, 34)

Defending her use of analogy, Thomas Pughe has argued, "We cannot speak about animals without using analogies- even if the idea of a union in suffering seems to deny this- and thereby we reveal the historical, and cultural, and moral conditions under which these analogies were produced" (2011, 11). The moral conviction that arguably motivates the consciousness of humans is based on a specific reason; however her argument, in this particular instance, is that to kill an animal is to be a soul-less human being, someone who lacks empathy. This political and ethical ambiguity has been termed by Pughe as "romantic primitivism." Nonetheless, Costello's defense becomes, in Huggan's words, "a fable of the impossible attempt to escape anthropocentrism" (2004, 713). Furthermore, Costello's own terms "sympathetic imagination" turns into a utopia beyond anthropocentrism and the commodification of nonhuman nature: "The image of nature survives because its complete negation by artefacts would necessarily involve closing one's eye to the possibility of a sphere beyond bourgeois work and commodity relations" (Adorno 1984, 102). To overcome Costello's contradictions, we have to keep in mind the relation of "significant otherness" (Haraway 2003, 3) between humans and animals: "The animal's is an otherness that keeps signifying ourselves and is hence, multiple, constantly productive and changing through the course of history and culture" (Barney 2004, 14). The questions asked by the text include the question of how to tell the story of humans and animals in such a way that the postcolonial world acknowledges the deprivations of colonial past and the power exerted by the global superpowers with deep colonial history. The novel's stance may appear to be ambiguous at a first glance; however, it becomes imperative that a writer should speak for animals from their position seems particularly relevant: "The responsibility of the writer to animals would thus consist in the attempt to put into language what their otherness signifies to us, the human animals, and thus to make new forms of expression emerge" (Pughe 2011, 15).

A counterpart to the philosophical *The Lives of Animals*, *Disgrace* again shows the author's preoccupation with the idea of the other, and how to transgress the culturally produced binaries between human and nonhuman, and tries, to some extent, to destabilize the same comfortable boundaries. The corporeal significance of dogs in a South African novel certainly could be read as carrying a deeply symbolic message; most critics have branded the novel as being overly pessimistic, however. Dionne Brand goes as far as to remark that the novel is incapable of ushering in any possibility of change because "Coetzee doesn't offer any other choices than death" (2002, 131). The focus on the philosophical arguments of the novel succeeds, more often than not, to overshadow the engagement with what the anthropologist Agustin Fuentes has called "the physiological and biostructural homologies" (2008, 124) between humans and other species. John Simons (2002) has argued that the nonhuman experience cannot be reproduced, only represented; hence, the narrative offered by Coetzee could only go as far as being symbolic, and works as a mode of questioning the existing all manners of categories, including racism, and speciesism that still pervades the Western imagination.

In *Disgrace* (2000), the distinction between philosophical animals and poetic animals, bulk of which dominated the narrative of Elizabeth Costello's argument, is far more nuanced and problematized but portrayed skillfully through the gradual transformation of David Lurie into being or rather becoming one of the dogs. The transformation is indeed surprising since the man so "corroded with skepticism" (Coetzee 2000, 102) was quite indifferent to animals and their welfare before moving out of Cape Town. Generally speaking, as is the case with any transformation, David's transformation too was far from being voluntary and sudden, and it took place only after his philosophical outlooks toward animals were threatened with bankruptcy after encountering the everyday animals in his daughter's life, and especially after he started working as a volunteer in her friend Bev's animal shelter; this second life parallels his former comfortable white academic life. David's descend into disgrace, into the life of animals, is far from that of the "sexy jaguars and huggable pandas" (Coetzee, 1999, 55) that the canon of literature prides itself on representing, but is one among the "cats . . . and dogs: the old, the blind, the halt, the crippled, the maimed" (Coetzee 2000, 218). He has become a "dog man: a dog undertaker, a dog psychopomp; a harijan" (Coetzee 2000, 146). Interesting here is Coetzee's use of the term "harijan" alongside "dog" indicating the plight of the people belonging from the lower castes in India, and how they have been perceived as for centuries, while at the same time also highlighting the fact that David now belongs to the margins with the not-so-sexy animals. The docile meek creatures seem to be utterly helpless, and at face value, they seem to lack the powerful aura of animals that could accelerate change but, as articulated by Tom Herron, "it is precisely as a consequence of their lack of power that they come to assume an exemplary, transformative status" (2005, 7). This perhaps echoes John Simons's claim that "animals can release love in human beings and thus transform them" (2002, 168).

Almost on every page of the novel, there could be found the mention of an animal. Lucy looks after white people's watchdogs trained to "snarl at the mere smell of a black man" (Coetzee 2000, 110) and used as deterrence in post-apartheid Africa. But not every animal mentioned in the novel is the white man's weapon against the blacks. The subservient animals "do not own their own lives" (Coetzee 2000, 123), and meet the fate of destitute in a world where they are utterly dependent on their human masters: "The lives of animals are routinely erased by human beings, not just through acts of physical violence but also by means of their exploitation in cliché, in pseudobioethical assertion, and in metaphor" (Herron 2005, 8). The "sympathetic imagination" of Costello finds its best expression through Lucy in *Disgrace*, and her love for animals acts as a foil to the utterly pessimistic outlook toward animals. She resonates much of what Costello tries her best to convey to her audience: "They do us the honor of treating us like gods, and we respond by treating them like things" (Coetzee 2000, 78). Donna Haraway in *The Companion Species Manifesto: Dogs, People, and Significant Otherness* has argued similarly that "[dogs] are not an alibi for other themes: dogs are fleshly material-semiotic presences in the body of technoscience. Dogs are not surrogates for theory; they are not here just to think with. They are here to live with" (2003, 5). The dogs or even the other animals featured in the novel is not the answer that could mollify the stained human soul; they are not here to offer salvation. The novel is not an advocate of the rights that animals deserve but represents the "white dilemma" that emerged in the post-apartheid South Africa, and which, to some extent, finds itself exposed through the animals.

The transgression of David into becoming more and more like a dog roughly starts in "Chapter 11" where he answers Lucy's queries with, "Why? Do you need a new dog-man" (Coetzee 2000, 88)? On the surface, even though he is talking about the actual job of a dog caretaker, syntactically, it may refer to the possibility of David becoming a "dog-man" with the key characteristics of the animal. This transformation does not entail a change in physical appearance but indicates, in Costello's term, the character's sympathetic imagination. Let us circle back to the first few chapters where David's identification with dogs is problematically expressed when he compares the dog's desire, which is instinctual to his own and describes himself as being "a servant of Eros" (Coetzee 2000, 52) in defense of the rape allegation against him. As ridiculous as that sounds, David's character resonates and identifies with that of the dogs in multiple ways, and this identification has very little to do with the dog's instinctual desire. What aligns David with the dogs is their shared suffering, a suffering that cannot be communicated through language, not only because it is impossible to move beyond the boundaries imposed by language between humans and nonhuman animals, but also because the suffering is deeply personal, something that is almost impossible to communicate even to someone belonging from same species; the experience of suffering could only be felt and expressed through solidarity in silence, through the unspoken words. David's sympathy that he extends to animals has a paradoxical dimension, and his views on animals fluctuate between being utterly indifferent and dismissive of their existence to being curious and genuinely invested in their future. The disdain that David feels is expressed in his conversation with Lucy when he replies: "Yes, I agree, this is the only life there is. As for animals, by all means, let us be kind to them. But let us not lose perspective. We are of a different order of creation from the animals. Not higher, necessarily, just different. So if we are going to be kind, let it be out of simple generosity, not because we feel guilty or fear retribution" (Coetzee 2000, 74). Interestingly, he describes his rendezvous with Soraya, the prostitute he visits every Thursday evening, as snakes copulating. Time and again he compares himself to predators and parasites. After he moves out to what he calls the "dark Africa" (Coetzee 2000, 95) that he grasps the extent to which animals are both present in his life, and will play a central role in his transformation also. An important episode that plays a major role in David's transformation is when a black man brings in his goat whose scrotum has been wounded by dogs. A male goat wounded in his scrotum and facing death almost becomes the animal double of David symbolically. Later after the horrific attack and Lucy's rape, when Bev tends to David's wound, he remembers the way she stroked the goat, and "to his own surprise" he finds himself saying: "Perhaps he has already been through it. Born with foreknowledge, so to speak. This is Africa, after all. There have been goats here since the beginning of time. They don't have to be told what steel is for, and fire. They know how death comes to a goat. They are born prepared" (Coetzee 2000, 83–84). It is obvious that he is comforting himself here, and a little later in what may be termed as the first exchange that David has with Bev without contempt, he first recognizes and voices his connection with other animals, between human and the nonhuman other: "'Do you know why my daughter sent me to you?' 'She told me you were in trouble.' 'Not just in trouble. In what I suppose one would call disgrace'" (Coetzee 2000, 85). Chapter 11 has been much discussed, and I would like to add that before the tall man shot the dog dead, Lurie's insight, "He speaks Italian, he speaks French, but Italian and French will not save him here in darkest Africa" (Coetzee 2000, 95), at once detaches him from his beliefs, and the shooting played a part in dehumanizing the men and thereby, in turn, humanizing the animals, and it gives us the readers the emotional relation that makes us a part of David's transformation.

Becoming an animal, the other does not consist in merely imitating an animal; becoming produces nothing other than itself. David's journey into becoming the animal-other is an irreversible journey with no way back. In *Kafka: Toward a Minor Literature* Deleuze and Guattari suggests:

> To become animal is to participate in movement, to stake out the path of escape in all its positivity, to cross a threshold, to reach a continuum of intensities where all forms come undone, as do all significations, signifiers, and signifieds, to the benefit of an unformatted matter of deterritorialized flux, of non-signifying signs. (1986, 13)

Becoming something other than himself occurs when David steps in Katy the bulldog bitch's cage, and stretching out beside her, he fast falls asleep. David's process of transmutation is not completely rid of his dismissal for animals, something that is a product of his lifelong held ideologies. However, his quick dismissal is punctuated, more often than not, with questions or even moments of genuine inquiry. In response to Bev's "They can smell what you are thinking" all David mutters to himself is "what nonsense!" (Coetzee 2000, 81). These types of dismissals are punctuated with queries such as "Who does the neutering?" (Coetzee 2000, 84), and "Are they all going to die?" (Coetzee 2000, 85). The time spent amid the animals, with them, an experience that is sometimes marked by a shared silent exchange of reciprocity.

The more David spends time in the "darkest Africa" the more he finds English unfit for articulating the truth for South Africa. He contends, "Like a dinosaur expiring and settling in the mud, the language has stiffened" (Coetzee 2000, 117). Lucy's decision to stay on the farm after the attack and rape as Petrus's extended family informs David of the changed power dynamic: "It is a new world they live in, he and Lucy and Petrus. Petrus knows it, and he knows it, and Petrus knows that he knows it" (Coetzee 2000, 117). Petrus, in David's eyes, is no longer just the gardener and the dog-man, he is from on considered a fellow human being, someone who reflects the complex history of his country. In a way, David will no longer be Lucy's protector and it is he who will become the "dog-man." The shifting power dynamic compels David to see the parallels between Lucy's rapists and himself as well as the connection that he shares with other animals. It is while he lives as a dog-man that the saving opera comes to him. David has the idea of introducing the voice of the dog who likes music affectionately called Driepoot by Bev. The three-legged dog mirrors David's own disgrace. David's acceptance of his own disgrace perhaps is best expressed when he refuses to grant this dog who "will have to submit to the needle" (Coetzee 2000, 215) one more week of grace. Driepoot, in a sense, is his muse for the opera: "The dog is fascinated by the sound of the banjo. When he strums the strings, the dog sits up, cocks its head, listens. When he hums Teresa's line, and the humming begins to swell with feeling . . . the dog smacks its lips and seems on the point of singing too or howling. Would he dare to do that: bring a dog into the piece" (Coetzee 2000, 215)? For David, the three-legged dog becomes more than just a metaphor for human dignity. "David carries the dog into the clinic building, into the theatre with its zinc-topped table where the rich, mixed smells still linger, including one he will not yet have met with in his life: the smell of expiration, the soft, short smell of the released soul" (Coetzee 2000, 219).

David makes the crossing of the boundary between humans and animals possible by attributing to the dying dog what had long been associated with human beings only: a living, breathing soul. The singing dog is the last animal to be euthanized in the novel, and the closing conversation that takes place between David and Bev posits the ethical point of the novel beautifully, almost poetically. The last line of the novel, "Yes, I am giving him up" (Coetzee 2000, 220) marks the complete transformation of David, completely decentered from the hierarchies of race, and in a way from speciesism "to live with nothing, not nothing but, but nothing, like a dog."

Coetzee's narratives, both the hybrid-discourse and the work of fiction, stand on the opposite side of the ecocriticism of the North, which has ignored the consequences of European colonialism. *The Lives of Animals* and *Disgrace* have successfully managed to offer a possible what Huggan (2004) termed "overlapping fields" between humans and nonhuman animals, which also takes into account the consequences of European expansion. Coetzee tries to work its way into the impossibility of speaking for the nonhuman nature, which obviously lack in human linguistic codes. Both these works oppose the imperatives of society that are so prevailing as to be regulated as hegemony; Costello especially points the readers toward what the modern world has come to terms with in their everyday lives. But to the extent that *The Lives of Animals* preaches the oppositional view, perhaps as a way out, it also points out the problems that grew out of the modern society with its values based on reason. Coetzee somewhat decenters the North and its ideologies, while at the same time retains some of its cultural practices, situates it within the context of an ex-colony, and assigns significance to its ecology, albeit without engaging strongly with the ecological discourse. Furthermore, his is an engagement with the animal "imagining alternative futures" (Huggan 2004, 721) especially through Lurie's transformation that questions the boundary between human and animals. The animals portrayed are the "everyday-things" that we perceive; they are part of both the domestic environment and the non-domesticated wild environment, making them neither a fully part of the human life nor completely outside of it. It is there, while at the same time it is not. The parallel that Coetzee draws between the animals and the humans and to the extent to which they influence the lives of the characters around them suggest that the characters too occupy a similar ambiguous position between being purely human and purely animal, at once calling the notion of purity, which has racial implications, to the witness box. I would like to conclude by stating that the two narratives, as Coetzee's 1993 Nobel citation suggests, "capture the divine spark" between the humans and the animals, as idealistic as it may sound.

BIBLIOGRAPHY

Adorno, Theodor W. 1984. *Aesthetic Theory*. Edited by Gretel Adorno and Rolf Tiedermann. London: Routledge.
Barney, Richard A. 2004. "Between Swift and Kafka: Coetzee's Elusive Fiction." *World Literature Today* 78 (1): 17–23. Accessed April 1, 2020. doi: 10.2307/40158351.
Bate, Jonathan. 2000. *The Song of the Earth*. Cambridge, MA: Harvard University Press.
Brand, Dionne. 2002. *A Map to the Door of No Return: Notes to Belonging*. 2001. Toronto: Vintage.
Coetzee, J. M. 1999. *The Lives of Animals*. Chichester: Princeton University Press.
———. 2000. *Disgrace*. London: Vintage.
Deleuze, Gilles, and Felix Guattari. 1986. *Kafka: Toward a Minor Literature*. Translated by Dana Polan. Minneapolis: University of Minnesota Press.
Foundation, Nobel. "James Maxwell Coetzee." October 2, 2003. https://www.nobelprize.org/prizes/literature/2003/press-release/ (accessed May 4, 2020).
Fuentes, Agustin. 2008. "The Humanity of Animals and the Animality of Humans: A View from Biological Anthropology Inspired by J.M. Coetzee's Elizabeth Costello." *American Anthropologist* 108 (1): 124–132.
Haraway, Donna. 2003. *The Companion Species Manifesto: Dogs, People and Significant Otherness*. Chicago: Prickly Paradigm.
Herron, Tom. 2005. "The Dog Man: Becoming Animal in Coetzee's "Disgrace"." *Twentieth Century Literature* 51 (4): 467–490. Accessed April 10, 2020. www.jstor.org/stable/20058782.
Huggan, Graham. 2004. "Greening Postcolonialism: Ecocritical Perspectives." *MFS Modern Fiction Studies* 50: 701–733.
McDonald, David A. 2002. *Environmental Justice in South Africa*. Athens, OH: University of Cape Town Press.
Parry, Benita. 1998. "Speech and Silence in the Fictions of J.M. Coetzee." In *Writing South Africa: Literature, Apartheid, and Democracy, 1970–1995*, edited by Derek Attridge and Rosemary Jolly, 149–165. New York: Cambridge University Press.
Pughe, Thomas. 2011. "The Politics of Form in J.M. Coetzee's "The Lives of Animals"." *Interdisciplinary Studies in Literature and Environment* 18 (2): 377–395. Accessed April 2, 2020. www.jstor.org/stable/44087706.
Shohat, Ella. 1992. "Notes on the "Post-Colonial"." *Social Text* 31/32: 99–113. Accessed April 17, 2020. doi: 10.2307/466220.
Simons, John. 2002. "The Animal as Symbol." In *Animal Rights and the Politics of Literary Representation*, edited by John Simons, 85–115. New York: Palgrave.

Skinner, Jonathan. 2003. "Ecopoetics." *WordPress*. Accessed April 1, 2020. https://ecopoetics.files.wordpress.com/2008/06/eco3.pdf.

Vital, Anthony. 2005. "Situating Ecology in Recent South African Fiction: J. M. Coetzee's "The Lives of Animals" and Zakes Mda's "The Heart of Redness"." *Journal of Southern African Studies* 31 (2): 297–313. Accessed April 8, 2020. www.jstor.org/stable/25064996.

Wolfe, Cary. 2003. *Animal Rites: American Culture, the Discourse of Species, and Posthumanist Theory*. Chicago: University of Chicago Press.

Worster, Donald. 1995. "Nature and the Disorder of History." In *Reinventing Nature? Responses to Postmodernist Deconstruction*, edited by Michael E. Soule and Gary Lease, 65–86. Washington, DC: Island Press.

Chapter 16

Provincializing Ecocriticism

Postcolonial Ecocritical Thoughts and Environmental-Historical Difference

Animesh Roy

Published in 1960, Chinua Achebe in his novel *No Longer at Ease* makes a wry gloss: "Let them come and see men and women and children who know how to live, whose joy of life has not yet been killed by those who claimed to teach other nations how to live" (1960, 45). Although Achebe's remark has usually been taken as a counternarrative to the broader orientalist discourse in which Africa was set up as a foil to Europe, challenging imperial representations of Africa by negation—by its absence of history, civilization, development, and so forth, it is difficult to ignore how the germ of postcolonial writing ecologically was implicit in the text. Achebe's novel through its fictional representation attempts to untangle how Western/Euro-American ecological vision suffers from an epistemological myopia as it is inflected by a sense of geographical parochialism and insularity, grossly unaware of the postcolonial realities of the environment, and often prioritizing an ecological historiography in which human history is relegated to moments of communion with nature and colonial history is relatively deemphasized. Achebe's fictionalization of the postcolonial ecology may be understood as a significant intellectual intervention to unsettle and write back against the eco-theoretical hegemony of a particular Western discourse—a discourse that has been projecting itself as the vanguard of planetary ecocritical consciousness, and imploring the postcolonial subjects toward postmaterialist values. However, Achebe's novel rather than proposing a distinct but alternative ecological vision was in line with a series of other African novels (here I may add with other postcolonial novels as well), which tried to uncover the vicious nexus existing between land—by extension, the earth and imperialism, and how imagination was central to the decolonizing process—a form of postcolonial ecological enquiry that predates modern Euro-American environmentalism of the 1960s and 1970s, and which essentially was not a derivative of it. Subsequent theoretical propositions by other postcolonial theorists deepened and extended such an ecological vision—a vision that hinted at how an understanding of the land was essential to an understanding of the historical process of colonialism. The Martinican author Franz Fanon and the Palestinian writer Edward Said argued how "land was a primary site of postcolonial recuperation, sustainability, and dignity" and how "imagination was vital to liberating land from the restrictions of colonialism [and] neo-colonial forms of globalization" (DeLoughrey and Handley 03). In his book *The Wretched of the Earth* (1961), Fanon observed how "for the colonized people the most essential value, because the most concrete, is first and foremost the land: the land which will bring them bread and, above all dignity" (1961, 9). Resonating much with Fanon and Michel Foucault, Edward Said in his *Orientalism* (1978) brought to the fore how colonial servitude begins with the loss of the land to the outsider by "having their land occupied, their internal affairs rigidly controlled, their blood and treasure put at the disposal of one or more Western power" (1978, 36). Taken together, such observations indeed attempted to defeat, dismantle, and thereby decolonize ecocriticism of the Euro-American epistemological hegemony, ridding it of the presumed centrality and normativity it has acquired, ignoring any other alternative voices in its peculiar subjectivity. Moving beyond the metropole to the colonies provided ecocriticism the much-needed space to accommodate a multiplicity of other alternative voices and to focus on the realities of the postcolonial ecologies and the environmental-historical differences.

This chapter attempts to develop a critique of mainstream Euro-American/Western ecological criticism of its foundational methodologies and focus primarily from the perspective of a sympathetic outsider, and instead proposes the need for ecological issues of the postcolonial countries to be adequately historicized. Such observations about Western/Euro-American ecocriticism are essentially historical, sociopolitical, and philosophical in nature. First, mainstream or Euro-American ecocriticism despite claiming for itself to be the vanguard of planetary ecocritical consciousness is heavily parochial to the extent of remaining obsessed primarily with environmental issues of Western Europe and North America, insulating itself from ecological consciousness of non-Western or postcolonial countries. Most of the ecocritical scholarships developed since its inception in the 1990s focused primarily either on British romanticism or American nature writings to such an extent that ecocriticism in fact ran the risk of being regarded entirely as an offshoot of British and American studies. Even contemporary ecocritical scholarship is heavily skewed toward the exploration of Western ecological texts and ideas; non-Western ecocritical enquiry is often assumed to be incidental and tertiary. Second, mainstream ecocriticism is predominantly a white movement and it has failed to voice for the environmental concerns of the non-whites or other ethnic groups. Cheryll Glotfelty anticipated such an observation in her book *The Ecocriticism Reader* (1996) when she enquired, "Where are the other voices?" She argued that "ecocriticism has been predominantly a white movement. It will become a multi-ethnic movement when stronger connections are made between the environment and issues of social justice, and when a diversity of voices are encouraged to contribute to the discussion" (1996, xxv). Similarly, Lawrence Buell in *Writing for an Endangered World* (2001) observed, "No treatment of environmental imagination can claim to be comprehensive without taking account of the full range of historic landscapes, landscape genres, and environmental(ist) discourses" (2001, 8). Despite such sympathetic observations by both Glotfelty and Buell, ecocriticism has primarily been a monolithic theoretical framework that has failed to address the heterogeneity of voices arising out of different races, classes, and ethnicities. Even when there have been certain sporadic concerted attempts to make ecocriticism multiethnic by accommodating a plurality of many different voices, it has limited itself focusing on issues of the ethnic groups primarily within the United States. Ethnic issues of the environment outside the national geographical border of the United States or Western Europe rarely found a voice within the mainstream ecocritical canon. Third, there is a perception among a certain section of ecocritics in the West that ecocriticism is peculiar to the rich nations of the West, a by-product of the move toward postmaterialist values in the West and interpretation of ecological consciousness as extant in postcolonial texts are at best a secondary new arrival on the ecocritical scene, thereby trying to marginalize and ignore the varied ecocritical voices emerging out of the global South or the postcolonial countries. Fourth, mainstream ecocriticism's overt reliance on the superficial aspects of nature has only served to concretize the split existing between nature and culture, that is, it got stuck in the same old binaries it purported to refute as postmodern theory. It failed to take into consideration societies where the relationship between the human and nature cannot be understood by simple nature/culture dichotomy, societies in which nature and culture are not different but co-constitutive. Fifth, Western ecocriticism in its ecophilosophical obsession for wilderness preservation and deep ecology has often ended up being reductive and xenophobic, and it failed to comprehend places where human and nature live in close proximity and harmony with each other; societies which have evolved locally specific and syncretic eco-cultural ethics. Sixth, Western ecocriticism is often hypocritical as it is preoccupied with issues of the environment within its own borders while being amnesiac of the ecological repercussions of the foreign policies of the West, particularly that on the non-Western countries.

Such observations stem from a broader understanding that the intellectual and geopolitical insularity of Western ecocriticism in engaging with issues of postcolonial ecologies originated out of the conceptual fatal flaw in conjuring humans as the single, transcendent villain responsible for the pernicious effect on planetary ecology—a flaw that shaped the narrative of an undifferentiated Anthropos amnesiac of the global power relations and thereby absolving the erstwhile colonial powers of the West of its complicity with the ecological crisis. Interestingly, ecocriticism being part of postmodern studies, it should have asserted the multiplicity of many different voices. Its disciplinary engagement with issues of ecology made it ideally suited for a planetary approach, much more transnational than any other field of literary and cultural enquiry. But ecocriticism limited itself dealing with ecological issues primarily that of the West. Again, theorizations of postcolonial ecologies tended to be shallow and reductive while trying to comprehend postcolonial ecological realities from a superior outsider position particularly with a Western gaze. Thus, a growing voice of dissent began to be heard from the postcolonial ecocritics and even a section of the ecocritics of the West who argued that the seemingly universal voice of ecocriticism did not represent the voice of the people of the postcolonial countries. They pointed out that there is a difference between the environmental spaces of the postcolonial countries with that of Walden or Wuthering Heights—ideal environmental spaces as conceived by the West. Rather the environmental realities of the postcolonial countries have to be adequately historicized and read in connection with the history of the Empire. This epistemological vacuity urged the need for ecocriticism to move beyond its self-enclosing limitations and diversify itself shunning its presumed normativity to accommodate a plurality of different ecocritical voices from the postcolonial world, and in the process making it much more capacious. However, there exist a mysterious silence between ecocritical and postcolonial studies. While postcolonial theory apparently remains silent on issues of the environment, a similar silence prevails with ecocriticism's engagement with postcolonial issues. In his essay, "Environmentalism and Postcolonialism" Rob Nixon has brought out four epistemological gaps existing between the dominant concerns of ecocritical and postcolonial theories. Nixon argued that while postcolonial theory is preoccupied with issues of hybridity and cross-culturation, ecocritical theory concerns itself with ideas of purity: virgin wilderness, pristine environment far from the social and political problems of the modern-day life. Second, while postcolonial theory deals with issues of migration, displacement, and globalization, ecocritical studies deal with ethics of place and belonging.

Further, while the postcolonial framework is critical of nationalism and favors a cosmopolitan and transnational discourse, literary environmentalism tends to deal with a parochial and national framework of deep ecology and wilderness preservation. Finally, if postcolonial theories try to exhume the marginalized colonized past, environmentalism tries to understand how most environmental history is repressed to solitary moments of communion with nature (2005, 235).

Despite such schisms existing between postcolonialism and ecocriticism as Nixon argued, a mutual engagement between the two epistemological fields is indeed essential to discern the environmental realities of the postcolonial countries and being amnesiac of the categories of race, class, gendered subaltern, and the history of colonial and postcolonial (here I may add neocolonial) servitude only serve to marginalize the rather long history of critique by eco-socialist, environmental justice activists, and writers who have repeatedly tried to expose the relationship existing between power, subjectivity, and place. Moreover, harnessing postcolonialism to ecocriticism lends sufficient rigor to literary and cultural analysis enabling to weld text and context together. Urging the need for a mutual dialogue between the two fields, Scot Slovic in his "Editor's Note" to the 2007 issue of *ISLE* pointed out that yoking the two disciplines would help to show "the value and necessity of this combination of perspectives" (2007, vi). Graham Huggan similarly argued that such an approach instead of being entirely new has rather "renewed, [and] belatedly discovered, its commitment to the environment" (2007, 702). In subsequent issues of *ISLE* particularly in 2007, Cara Cilano and Elizabeth DeLoughrey collected several papers on the postcolonial interpretation of ecocriticism from a wide number of sources ranging from the postcolonial ecologist Ramachandra Guha to ecofeminist works to argue that such a theoretical proposition though entirely new to a Western-based audience would help to "rethink the limitations of US national frameworks that had occluded other perspectives" (2007, 73). Voicing the urgent need for a postcolonial interrogation of the African-American landscapes, Christine Gerhardt in her paper "The Greening of African-American Landscapes: When Ecocriticism Meets Post-Colonial Theory" argued how the postcolonial perspective provides a critical tool to explore how black literature draws attention to "the ways in which the questions typically asked by ecocriticism need to be rephrased . . . particularly with regard to discussions of nature and race that do not participate in the very mechanisms of exclusion they are trying to dismantle" (2002, 516). Somewhat similarly, Byron Caminero-Santangelo linked African environmental literature to the politics of decolonization, a politics that might be overlooked if read from a strictly mainstream American ecocritical perspective. Upamanyu Pablo Mukherjee crucially explained how the two different fields instead of being two separate disciplines are mutually inclusive, for any field purporting to theorize the global conditions of colonialism and imperialism (let us call it postcolonial studies) "cannot but consider the complex interplay of environmental categories . . . with political or cultural categories, just as any field purporting to attach interpretative importance to environment . . . (let us call it eco/environmental studies) must be able to trace the social, historical and material co-ordinates of categories" (2010, 144).

Unlike mainstream ecocriticism, which led to homogenizing and totalizing of ecocritical theories because of a Western gaze, postcolonial ecocriticism focused at the points of intersection between ecological history and colonial history, thereby trying to emphasize how the two are mutually imbricated for, as DeLoughrey and Handley argued that "to deny colonial and environmental histories as mutually constitutive misses the central role the exploitation of natural resources plays in any imperial project" (2011, 10). Trying to explore the geographical ramifications of colonialism Edward Said explained how imperialism and violence toward the earth is intricately interlinked:

> Imperialism after all is an act of geographical violence through which virtually every space in the world is explored, charted, and brought under control. For the native, the history of colonial servitude is inaugurated by the loss of the locality to the outsider; its geographical identity must thereafter be searched for and somehow restored. (1994, 77)

Richard Grove too exposed the relationship existing between imperialism and environmental violence when he argued in his book *Green Imperialism: Colonial Expansion, Tropical Island Edens and the Origins of the Environmentalisms, 1600–1860* that "colonial experience was not only highly destructive in environmental terms but that its destructiveness had its roots in ideologically 'imperialist' attitudes towards the environment" (1995, 6). Interestingly, since the material resources of the colonies were important to the metropolitan center, some of the earliest environmental conservation practices were instituted in the colonies, which served the dual function of presenting the colonial power as benevolent on one hand and screening the imperialist and exploitative nature of the empire on the other. Richard Grove observed that, "early environmental concerns, and critiques of the impact of western economic forces on tropical environments, in particular, emerged as a corollary of, and in some sense as a contradiction to, the history of the mental and material colonization of the world by the Europeans" (1995, 2). Such environmental conservation practices had very little to do with the colonialists' love and preservation for the environment, rather it was devised to check the rapid environmental degradation of the colonies as a result of ruthless colonial expansion of the empire attended by a plantocratic society and rapid denudation of the forest for agriculture and shipbuilding industry. The Caribbean Islands is a glaring example of an endangered ecosystem resulting from natural resource exploitation by the empire. The Caribbean Islands is generally exoticized as an Edenic retreat for its tourism potential. However, its environmental history is a testimony of the fact that rapid environmental exploitation of the islands by the British for the sake of cash crop cultivation like sugarcane led to severe water depletion in the island, which, in turn, forced the island to be dependent solely on tourism. The history of ecological violence in the Caribbean is an example of what Lawrence Buell implies by "environmental doublespeak" (1997, 4), something that V. S. Naipaul brings out in his autobiography *The Enigma of Arrival* (1987). Quite similarly, Jamaica Kinkaid in *A Small Place* (1988) explores how the British colonial system denuded the entire forest of Antigua and replaced the local population who protected them with imported slaves for the creation of a plantocratic economy making the whole island dry and solely dependent on tourism, that is, it is the exploitation of the transatlantic colonies, which made the luxury and opulence of the imperial center possible. Rob Nixon's wry gloss that "ironically, a place scarred by a long history of coercive labor and violence has been reinvented as an Edenic retreat where Europeans and North Americans can experience nature as pure—a paradise beyond reach of work and time" essentially brings out the inherent duplicity implicit within Western ecodiscourse in presenting nature as pristine and paradisiacal, while occluding a much brutal history of severe ecological depredation (2005, 241). The British forest management in India is another example. The forest management service under the colonial British left the forests in India in a much worse state than it was before. Indeed, as Ramachandra Guha puts it (citing a Scottish forester): "Is it not the case that the history of civilized man in his colonization of new countries has been in every age substantially this—he has found the country a wilderness; he has cut down trees, and he has left it a desert" (2000, 28). This is how the genealogy of an otherwise benevolent Western environmentalism in postcolonial countries can be traced back to an earlier period of intensive environmental exploitation. DeLoughrey and Handley's observation is pertinent that "the environmental sciences that tell us that we can no longer afford to ignore our human impact on the globe are an ironic by-product of a global consciousness derived from a history of imperial exploitation of nature . . . . Thus the nostalgia for a lost Eden, an idealized space outside of human time, is closely connected to displacing the ways that colonial violence disrupted human ecologies" (2011, 2–13).

Colonial expansion and its consolidation also led to the promotion of an entirely patriarchal mode of Western discourse of scientific ideas of dominance and control—an overarching colonial design of power and control that becomes evident in Sir Ronald Ross's influential note "The Malaria

Expedition to West Africa" where he argued that "the success of imperialism would depend largely upon the success of the microscope" (1900, 36). There is ample scholarship in social and natural sciences, as DeLoughrey and Handley argue, to demonstrate that "Western discourses of nature and environment have been shaped by the history of the empire" (2011, 10). In the eighteenth century, the European mania for plant collection particularly that of the New World made possible the production of Carolus Linnaeus's binomial taxonomy. The European knowledge building exercise leading to the biotic reconfiguration through a radical remapping of the global space through the common language of Latin not only led to the creation of a hierarchy but also a homogenization of the entire plant species bereft of local cultural moorings. Michel Foucault argues that "for taxonomy to be possible . . . nature must be truly continuous, and in all its plenitude. Where language required the similarity of impressions, classification required the similarity of impressions, classification requires the principle of the smallest possible difference between them" (1994, 159). Linnaeus hierarchical model reinforced the idea that scientific knowledge is created by great individuals especially the Europeans, which by corollary led to the institution of hierarchy within humans, further contributing to a biological determinist discourse based on race, class, gender, and nature. This episteme of difference centered on the biological determinist discourse and supported by the enlightened theorists' argument of climatic determinism was used to validate the skewed vision that people of the tropics were inferior to and incapable of reaching the moral and cultural heights as that of the West; thereby, it was often the White Man's Last Burden to bring out the natives from such backwardness. This was how the knowledge apparatus of the empire was used to justify slavery, denial of citizenship, and subjectivity to the non-Europeans thereby consolidating the foundation of the empire further. Linnaeus's binomial nomenclature, mnemonic and honorific of elite Western botany as it was, detaching plants from their native cultural moorings has been critiqued by Jamaica Kinkaid in *My Garden*:

> These countries in Europe shared the same botany, more or less, but each place called the same thing by different names: and these people who make up Europe were (are) so contentious anyway, they would not have agreed to one system for all the plants they had in common, but these new plants from far away, like the people far away, had no history, no names, and so they could be given names. (1998, 122)

Thus, eighteenth-century science and knowledge production served as the tools of the empire swallowing and destroying the diverse geographic and cultural identities of the world flora, and instead established a reductionist knowledge base comprehensible first and foremost to the Europeans while discarding all others as insignificant and secondary.

While it can barely be doubted that the roots of the current environmental concerns in the postcolonial countries may be traced back to a period of intense environmental exploitation and European global expansion during colonial times it is also important to uncover the complex reticulate relationship existing between environmental degradation and neocolonial and neoliberal orders in postcolonial era. Though colonialism has formally come to an end with the end of World War II, informally it inaugurated a period of more intensive and sustained exploitation of the erstwhile colonies. Edward Said argued in *Culture and Imperialism* (1994) that colonialism did not come to an end, rather it changed its form, for "Westerners may have physically left their old colonies in Africa and Asia, but they retained them not only as markets but as locales on the ideological map over which they continue to rule morally and intellectually" (1995, 27). While during the colonial period, the ecologies of the former colonies were exploited by the metropolitan centers in search for certain material riches to fuel their Industrial Revolution, in postcolonial period, the colonies were turned into what Fanon calls "an economically dependent country" (2004, 55)—potential rich locales that could sustain their economy. With the end of the Cold War, the balance of global power shifted increasingly from the Europeans to the North Americans, chiefly the United States—a nation which has come to be regarded in recent times as the vanguard of modern environmentalism, but ironically, which "has done far less than one might reasonably expect to protect the global environment but far more than it could possibly have hoped to 'reinvent the imperial tradition for the twenty-first century' (Lazarus 2006, 20)—a country that has actively and aggressively contributed to what many now acknowledge to be the chronic endangerment of the contemporary late-capitalist world" (Huggan and Tiffin 2010, 1). The reinvention of the imperial tradition for the twenty-first century, as argued by Huggan and Tiffin, refers to "a historical condition of intensified and sustained exploitation of the majority of the humans and the non-humans of the former colonies by a cartel composed of their own and 'core' metropolitan European/north American elites . . . . The globalized ruling classes of this postcolonialism, whose interests are often embodied in gigantic transnational corporations and the labyrinthine world of speculative financial transactions, are often called the new cosmopolitan" (Mukherjee 2010, 6). Rob Nixon likens this to "an era of resurgent imperialism, an era in which—sometimes through outright, unregulated plunder, sometimes under camouflage of developmental agendas—a neoliberal order has widened, with ruinous environmental repercussions, the gulf between the expanding classes of the super-rich and our planet's 3 billion ultrapoor" (2011, 37). Although rarely referred to by American ecocritics, the "chronic endangerment" posed by the

United States to other countries derives from excessive consumption (even of nature as something that is consumed by the rich elites for their aesthetic pleasure), pollution and the neoliberal forms of globalization, militarization, and development. Linda Colley very pertinently remarks that "we may be living in post-colonial times, but we are not yet living in post-imperial times" ("What Is Imperial History Now" quoted in Nixon, *Slow Violence* 233).

One of the central tasks of postcolonial ecocriticism is to contest and not unoccasionally to provide an alternative to the Western ideology of development, progress, and globalization. Given the history of the international model of development and progress that landscapes and restructures national spaces, the people of the postcolonial nations do not have the luxury to be oblivious of crucial environmental pressures. The various developmental schemes initiated by the International Monetary Fund, World Wildlife Fund, and World Bank, such as keeping aside places for conservation projects, resource extraction, the use of agrichemical fertilizers and patented seeds, have all radically altered the environment of the non-West. Postcolonial critics see this model of development "as little more than a disguised form of neo-colonialism, a vast technocratic apparatus designed to serve the interest of the West" (Huggan and Tiffin 2010, 27). Wolfgang Sachs so pertinently argues that "what development means depends on what the rich nations feel" (1997, 26). This myth of development, as DeRivero argues, taking false support from the Western Enlightenment ideology of progress and the socio-Darwinian principle of the survival of the fittest, enjoins the less "advanced" Southern nations to close the gap with their wealthier counterparts, and in doing so subscribe to a capitalist model that is not only unequal but also with devastating environmental costs. Vandana Shiva has rightly indicated that in case of the non-Western/ postcolonial countries, development has been coterminous with industrialization, and that "the dominant model of development and globalization is inherently violent because it deprives the poor of their fundamental right to food, land and livelihoods" (1999, 4). Critiquing India's postcolonial agenda of development especially how the construction of magadams was pushed as the supreme emblem of "National Progress," as "Temples of Modern India" (Roy 2002, 40) and people were asked to "suffer in the interest of the country," Arundhati Roy argued: "How can you measure Progress when you don't know what it costs and who has paid for it? How can the market put a price on things—food, clothes, running water—when it doesn't take into account the real cost of production?" (2002, 43). Roy went on to argue that the multi-billion-dollar dam industry having been decommissioned in the West for its potential ecological and social threats has been deliberately pushed in the non-Western countries in the name of development and progress. Roy finds such policies "undemocratic" and a means of looting poor people's land, water, and irrigation and giving it to the rich (2002, 42). While globalization should have ideally meant a global transfer and accessibility of trade and ideas between and across nations, "present day globalization," as DeRivero argues, "is the result not so much of free global competition among nations, but of a network of agreements and productive and financial activities among the transnational corporations" (2001, 29). DeRivero stresses that in contrast to the erstwhile colonial powers who often tried to balance their national ambitions with their international responsibilities such as the protection of human rights and environmental degradation, the executives of the transnational companies, going by neoliberal policies, do not want to establish any link between their global negotiations and the devastating environmental problems they cause (2001, 33). Helon Habila's *Oil on Water* (2010) and Indra Sinha's *Animal's People* (2007) presents vivid fictional representations of how the Western narrative of globalization proves counterproductive to the ecology and the people in the non-West. Both Habila and Sinha fictionalize a chilling truth, about how forces of globalization transform the material realities of the non-West/postcolonial countries when transnational companies in luring after the benefits of globalization outsource toxicity in the non-West where labor is cheap and new markets to tap. The omnipresence of sites of intensified environmental exploitation in the non-West intricately linked to the ill effects of globalization is brought out by Sinha when he fictionalizes Khaufpur as the real-life Bhopal—the site for world's worst industrial disaster: "Is Khaufpur the only poisoned city? It is not. There are others and each one of has its own Zafar. There'll be a Zafar in Mexico City and others in Hanoi and Manila and Halabja and there are the Zafars of Minamata and Sevaso, of Sao Paolo and Toulouse" (2007, 296). The concern for the ecologically devastating effects of petroculture in the Niger Delta as voiced by Habila has much earlier been pointed out by Ken Saro-Wiwa—generally regarded as Africa's first environmental martyr. But his writings hardly found a place in Western ecocritical canon perhaps because he was exploring the vicious links existing between ethnicity, pollution, human rights, and the transnational corporations rather than sticking to the Western ecophilosophical models propounded by Thoreau and Jefferson. A similar sentiment is expresses by Arundhati Roy when she argued that globalization in the postcolonial countries not only leads to depredation of the ecology but also the disenfranchisement of the socially marginalized people who depend on it: "I think of globalization like a light which shines brighter and brighter on a few people and the rest are in darkness, wiped out. They simply can't be seen. Once you get used to not seeing something, then, slowly, it's no longer possible to see" (quoted in Nixon, *Slow Violence and the Environmentalism of the Poor* 2011, 1).

Postcolonial ecocriticism has come to explore the environmental repercussions of the American foreign policy, especially the transnational fallout of the American environmental practices, the disproportionate impact of the U.S. global policies and ambitions that have had and are having a serious effect on socio-environmental landscapes internationally. Rob Nixon argues that the U.S. foreign policies have had a long history of reinforcing "asymmetrical relations between a domestically regulated environment and unregulated environment abroad" (2011, 35). This phenomenon that Nixon terms as "superpower parochialism" means "a combination of American insularity and America's power as the preeminent empire of the neoliberal age to rupture the lives and ecosystems of non-Americans, especially the poor, who may live at a geographical remove but who remain intimately vulnerable to the force fields of U.S. foreign policy" (2011, 34). Robert Barclay's *Melal: A Novel of the Pacific* (2002) is a fictionalization of superpower parochialism at work as it explores the unimaginable environmental horrors faced by the people of the Marshall Islands as a result of continuous nuclear testing in the Pacific by the United States. Barclay's text written in the background of the Bravo test of 1954 demonstrates how the United States while protecting its own citizens from the radioactive fallout of nuclear tests were least concerned about the ecological and health hazards of people and the Marshall Islands. A corollary to this "superpower parochialism" is the environmental racism implicit within the U.S. foreign policy. Environmental racism, defined by the American environmental philosopher Deane Curtin, is "the connection, in theory and practice, of race and the environment so that the oppression of one is connected to, and supported by, the oppression of the other" (*Environmental Ethics for a Postcolonial World* 2005, 145, as quoted in Huggan and Tiffin 2010, 4). Huggan and Tiffin defines environmental racism as:

> a sociological phenomenon, exemplified in the environmentally discriminatory treatment of socially marginalized or economically disadvantaged peoples, and in the transference of ecological problems from the "home" to the "foreign" outlet (whether *discursively*, e.g. through the more or less wholly imagined perception of other people's "dirty habits," or *materially*, e.g. through the actual rerouting of First World commercial waste.). (2010, 4)

The meticulously planned nuclear testing by the United States may be understood as a form of casual environmental racism indulged in toward the indigenous inhabitants of the region, where the indigenous Marshallese people were treated as mere guinea pigs so that "the Americans could test what happened to people as a result of their bombs" (Barclay 2002, 81). Seeking "empty" test spaces, the United States decided to carry on with its Cold War program of nuclear testing in the Marshall Islands. But the island region was not *terra nullius* but inhabited by indigenous peoples for thousands of years. Such an idea of environmental racism is implicit in the words of Lawrence Summers, the president of the World Bank, who in 1991 argued for the dirty industries to be shifted to non-West or postcolonial countries:

> I think the economic logic behind dumping a load of toxic waste in the lowest-wage country is impeccable and we should face up to that...I've always thought that countries in Africa are vastly under polluted; their air quality is probably vastly inefficiently low compared to Los Angeles . . . . Just between you and me, shouldn't the World Bank be encouraging more migration of the dirty industries to the Least Developed Countries? (Confidential World Bank memo, December12, 1991, as quoted in Nixon, *Slow Violence and the Environmentalism of the Poor* 2011, 1)

Lawrence Summer's argument for the migration of the dirty industries to the least developed countries is in line with the economic logic of neoliberalism where transnational corporations try to internalize profits and externalize risks. Summers, through his toxic distribution argument doubly discounted the peoples of the postcolonial countries: as bereft of an ecological concern of their own and even human entities.

Postcolonial ecocriticism thus instead of sticking to the Western hegemonic Thoreauvian or Jeffersonian ecophilosophy presses for a more comprehensive and cosmopolitan ways of approaching dominant Western ecodiscourse by accommodating those provincial voices, which were generally regarded as insignificant in its peculiar subjectivity. It attempts to address the "intellectual challenge" of drawing "on the strengths of bioregionalism without succumbing to ecoparochialism" that would "render ecocriticism more accommodating of what [Nixon calls] a transnational ethics of place" (2005, 239). This approach may not be "green" in the traditional ecocritical sense of the term, but perhaps it would go some way in bringing what Lawrence Buell describes as engaging " 'green' and 'brown' landscapes, in conversation with each other" (2001, 7). However, just as traditional ecocriticism's overt reliance on the philosophies of wilderness preservation and deep ecology denying any alternative definition of wilderness and environment limits its effectiveness, so too does the activist orientation of the environmental justice movement inherent within postcolonial ecocriticism limit its reach. A comprehensive and holistic engagement between the two, instead, would make ecocritical theory much more capacious.

BIBLIOGRAPHY

Achebe, Chinua. 1960. *No Longer at Ease*. Johannesburg: Heinemann.
Barclay, Robert. 2002. *Melal: A Novel of the Pacific*. Honolulu: University of Hawai'i Press.
Buell, Lawrence. 1997. *The Environmental Imagination: Thoreau, Nature Writing and the Formation of American Culture*. Cambridge, MA, London: Harvard University Press.
———. 2001. *Writing for an Endangered World: Literature, Culture and the U.S and Beyond*. Cambridge, MA, London: Harvard University Press.
Caminero-Santangelo, Byron. 2007. "Different Shades of Green: Ecocriticism and African Literature." In *African Literature: Anthology of Theory and Criticism*, edited by Tejumola Olaniyan and Ato Quayson, 698–706. Malden, MA: Blackwell.
Cilano, Cara, and Elizabeth DeLoughrey. 2007. "Against Authenticity: Global Knowledges and Postcolonial Ecocriticism." *ISLE* 14 (1): 71–87. JSTOR.
DeLoughrey, Elizabeth, and George B. Handley. 2011. "Introduction." In *Postcolonial Ecologies: Literatures of the Environment*, edited by Elizabeth DeLoughrey and George B. Handley, 3–39. New York: Oxford University Press.
DeRivero, Oswaldo. 2001. *The Myth of Development: The Non-Viable Economies of the Twenty-First Century*. London and New York: Zed Books.
Fanon, Franz. 2004. *The Wretched of the Earth*. Translated by Richard Philcox. New York: Grove Press.
Foucault, Michel. 1994. *The Order of Things: An Archaeology of the Human Sciences*. New York: Vintage Books.
Gerhardt, Christine. 2002. "The Greening of African-American Landscapes: When Ecocriticism Meets Postcolonial Theory." *The Mississippi Quarterly* 55 (4): 515–533. JSTOR.
Glotfelty, Cheryll. 1996. "Introduction: Literary Studies in an Age of Environmental Crisis." In *The Ecocriticism Reader: Landmarks in Literary Ecology*, edited by Cheryll Glotfelty and Harold Fromm, xv–xxxiii. Athens and London: The University of Georgia Press.
Grove, Richard. 1995. *Green Imperialism: Colonial Expansion, Tropical Island Edens and the Origins of the Environmentalisms, 1600–1860*. Cambridge: Cambridge University Press.
Guha, Ramachandra. 2000. *Environmentalism: A Global History*. New Delhi: Oxford University Press.
Huggan, Graham. 2004. "Greening Postcolonialism: Ecocritical Perspectives." *Modern Fiction Studies* 50 (3): 701–733. JSTOR.
Huggan, Graham, and Helen Tiffin. 2010. *Postcolonial Ecocriticism: Literature, Animals and Environment*. London and New York: Routledge.
Kinkaid, Jamaica. 1998. *My Garden*. New York: Farrar, Straus and Giroux.
Mukherjee, Upamanyu Pablo. 2010. *Postcolonial Environments: Nature, Culture and the Contemporary Indian Novel in English*. New York: Palgrave Macmillan.
Nixon, Rob. 2005. "Environmentalism and Postcolonialism." In *Postcolonial Studies and Beyond*, edited by Ania Loomba, et al., 233–251. Ranikhet: Permanent Black.

———. 2011. *Slow Violence and the Environmentalism of the Poor*. Cambridge: Harvard University Press.
Ross, Sir Ronald. 1900. "The Malaria Expedition to West Africa." *Science* 11 (262): 36–37. JSTOR.
Roy, Arundhati. 2002. *The Algebra of Infinite Justice*. London: Penguin Books.
Sachs, Wolfgang (ed.). 1997. *The Development Dictionary: A Guide to Knowledge as Power*. London: Zed Books.
Said, Edward W. 1994. *Culture and Imperialism*. London: Vintage Books.
———. 1995. *Orientalism: Oriental Conceptions of the Orient*. New Delhi: Penguin Books.
Shiva, Vandana. 1999. *Staying Alive: Women, Ecology and Development*. New York and Boston: South End Press.
Sinha, Indra. 2007. *Animal's People*. London: Pocket Books.
Slovic, Scott. 2007. "Editor's Note." *ISLE* 14 (1): v–vii. JSTOR.

Index

Africa, 6–7, 47–58, 184–89, 211
Alfred Crosby, 117
animism, 42, 140–44, 172–77. *See also* creatureliness; transcorporeality
Anthony Carrigan, 12, 62, 142
anthropocentric ecocriticism: Anthropocene, 118, 141, 145–47; anthropocentricism, 117, 140–51, 197, 203
architectural determinism, 156–58, 168
August Town, 156–60

bare life, 172, 178–79
bioethics, 9
built environment, 157–58

Canada, 89–91, 93–97
capitalism, 95–97, 132–41, 183–96
Caribbean islands, 128, 155, 217
carnival, 166
Cheryll Glotfelty, 119, 141, 213
Christopher Manes, 3
circuit of culture, 124
coloniality, 99, 139–40, 151. *See also* colonial products
colonial products, 9
contemporary conservation movement, 43
creatureliness, 179

cultural nationalism, 31–33, 44–45

deep ecology, 141, 213–15, 223
dislocation, 51, 58, 76, 83. *See also* forced migration; forced relocation
Donna Haraway, 205

ecocinema, 119, 124. *See also* ecofilm
ecocritical postcolonial studies, 156
ecofilm, 115
ecological imperialism, 117
Edward said, 72, 116, 139, 155, 165, 212–19
Elizabeth DeLoughrey, 9–11, 142, 215
enclosure, 10. *See also* enclosure act
enclosure act, 10
environmental injustices, 128, 137. *See also* environmental racism
environmentalism, 19–20, 63–64, 90, 197–98, 212. *See also* Euro-American environmentalism
environmental racism, 11, 118, 222–23
Euro-American environmentalism, 212
Eurocentricism, 140
European empire, 116, 201

Felix Guattari, 123, 202, 207
forced migration, 137. *See also* forced relocation

forced relocation, 132, 137
Franz Fanon, 116, 139, 155, 212, 219

Gayatri Spivak, 139, 165
Graham Huggan, 11, 141, 150, 215
green orientalism, 11
green postcolonialism, 155, 159–60, 163–64
griot, 190

habitual configuration, 198
Helen tiffin, 11, 141, 150

imperialism, 11, 25, 62, 84, 115, 183, 216. *See also* ecological imperialism
indigenous peoples, 128–29, 137, 223; jibara, 132, 136; jibaro, 129, 133
Indo-Pak subcontinent, 76, 84–85

jumbie, 163–65

Kafka's "A Report to an Academy", 202
Kerala, 33–46; advent of Buddhism to Kerala, 35; *Aithihyamala*, 37–41

Lawrence Buell, 213, 217, 223
Lynn White, 3

Mau Mau, 190, 193
media theorists, 155–56
Michael Fisher, 2, 4
mimicry, 91, 93, 155
monoculture colony, 131

neocolonialism, 90–95, 142, 190, 220
neoliberal, 219–23
north east, 63–69, 73

oil war, 50, 54–55. *See also* petrol imperialism

overlapping, 208

petrol imperialism, 51
planetarity, 119, 123
postcolonial environment, 141, 164
postcolonial India, 31–33, 104
postcolonial theories, 156–68, 214–15
pro-poor tourism, 12

Ramachandra Guha, 215, 217
Robert P. Marzec, 9–10
Rob Nixon, 63, 214–15, 219, 222

sacred groves/*kaavus*, 31, 34–39, 41–45; grove entrepreneurship, 45
settler colonialism theory, 128, 137; settler colonialism, 127–29, 132, 137
shared subjectivity, 177, 181
Shashi Tharoor, 4
slatees, 186. *See also* traitees
slave trade, 46, 186
South America, 3, 128–29, 199
sovereign subject, 171–78
speciesism, 140, 200–208
stock piling, 10
sympathetic imagination, 201–5

Timothy Clark, 62
tourism, 11–12, 193, 217. *See also* pro-poor tourism
traitees, 195
transcorporeality, 176
transmutation, 207

Urhobo, 51
utopia, 17–25, 27–28

Val Plumwood, 118
Vandana Shiva, 220

Walter Rodney, 1, 7, 184

About the Contributors

Anik Sarkar is a PhD research scholar in English at the University of North Bengal. He is an assistant professor of English at Salesian College, Siliguri. His latest publication is a chapter "Subjugation and Resilience in the Poems of Meena Kandasamy" in an edited volume, *Perspectives on Indian Dalit Literature* (2020), edited by Dipak Giri. He also writes fiction and his story *The Man Who Sold Diseases* was published by Juggernaut (2018).

Animesh Roy is currently working as an assistant professor at the Department of English, St. Xavier's College, Simdega in Jharkhand, India. His areas of research interests are environmental humanities, energy humanities, Petrofiction, religion and ecocriticism, climate fiction, ecocriticism and the Global South, and the North-South discourse. His publications are *Ecology, Literature, and Culture: An Anthology of Recent Studies* (2020) and *Ethnomedicine: Indigenous Culture, Plants, and Medicine* (2020) from Atlantic Publishers.

Anupama Nayar is working as a faculty in the Department of English and director of Centre for Concept Design at CHRIST (Deemed to be University), Bangalore. She has worked in the area of Joycean Studies for her doctoral thesis. She has a postgraduate diploma in English studies from CIEFL (Central Institute of English and Foreign Languages) and personnel and human resources management from LIBA (Loyola Institute of Business Administration). Her eighteen years of experience spans teaching, counseling, material production, creative writing, research guidance, and publication in the academic field. Her other areas of interest include postcolonial cultural studies, precarity studies, pedagogic studies, and organizational behavior and culture. She has national and international publications to her credit. She offers papers in British literature, postcolonial literatures and creative writing at the MA level. She also regularly conducts workshops on digital learning and content development. Also, she is a cultural history aficionado and uses creative writing as a stress buster. Short stories are her forte. Teaching is her passion.

Bipasha Mandal received a master's degree in English literature from Jadavpur University and a bachelor's degree from Calcutta University. Her research interests include postcolonial studies, marginality studies, Indian writing in English, and ecocriticism. Her work is published in several journals, conferences, and books.

Chinmaya Lal Thakur is a doctoral researcher in the Department of Creative Arts and English at La Trobe University, Melbourne. His work concerns the representation of subjectivity in the novels of David Malouf. Postcolonial studies, novel theory, literary criticism, modernist literatures, and continental philosophy constitute the areas of his interest. Apart from having published a number of essays and critical reviews in reputed national and international journals, he has also edited *Literary Criticism: An Introductory Reader* for Worldview Publications, New Delhi and Mauritius.

Denise M. Jarrett is assistant professor, specializing in Caribbean Literature in the English and Language Arts Department at Morgan State University, Baltimore, Maryland. She also has special interest in Postcolonial Literatures and Ethnic and Cultural Studies. Her publications include: "Identity Development and Survival Strategies in Selected Novels by Michael Anthony and Cyril Everard Palmer," "Carnival and Southern Games: National Tradition and Personal Identity in Michael Anthony's *The Games Were Coming*," and "Reading 'Black' Poverty in Postcolonial Caribbean Young Adult Fiction: Michael Anthony's *The Year in San Fernando* and Cyril Everard Palmer's *The Cloud with the Silver Lining*." She has also presented numerous scholarly research papers at the College Language Association Convention and West India Literature Conference.

Humaira Riaz is presently working as an assistant professor in the Department of English in City University of Science & Information Technology, Peshawar, Pakistan. She has completed her PhD in English literature. Title of her doctoral thesis was "Racism and Islamophobia: A Critique of Selected American Literary Texts." Her published work can be accessed and followed on www.researchgate.net & Academia.edu. Her areas of interest are racism, postcolonial literature, ecocriticism, feminism, and gender studies.

Kalpana Bora Barman is an assistant professor in the Department of English, Cotton University, India. She engages with postcolonial literature, the interstices of gender and literature, and writings from India's north east. She is particularly interested in narratives of the urban and has been exploring the complexities of the urban-spatial intertextualities and earning a doctorate for the same from the Indian Institute of Technology Guwahati, India. Her publications include "'Maximum City': Mumbai, Spatial Politics and Representation" in journal *Space and Flows: An International Journal of Urban and Extra Urban Studies*, Vol. 1.3 (2011), " 'A Journey in the Footsteps of Xuanxang': Mishi Saran's *Chasing the Monk's Shadow*" in *Negotiations*: *A Peer Reviewed Journal of the Department of English, Cotton College*, Vol. 3 (2015); and "The Spaces of the Everyday: A Reading of Temsula Ao's *Laburnum for My Head*" in *Innerscape; Exploring Literatures from India's North East* (2019).

M. Anjum Khan works as an assistant professor of English in Avinashilingam Institute for Home Science and Higher Education for Women, Coimbatore. Her areas of research have been British literature, immigrant Canadian literature, and cultural literary theories. Also, she is interested in teaching subjects like history, literature, disability studies, and literary theories. She is author of *Ethnic Silhouettes: M.G. Vassanji in the Light of New Historicism* and *Narrating Bodies, Reading Anosh Irani*. Moreover, she has published several research articles in reputed national and international journals and chapters in books, presented papers in national and international conferences, and has conducted workshops on journalism and assistive technology. She also delivers motivational speeches in colleges and corporate institutes.

Neepa Sarkar is an assistant professor in the Department of English, Mount Carmel College, Autonomous, Bangalore, and has completed her doctoral thesis on literature and collective memory. She won the Issac Sequiera Memorial Award for the Best Paper in the 17th MELOW conference (2018) and has published research articles in international journals including *History Today*, *MeJo Journal*, *Journal of Literature and Aesthetics*, *Global Colloquies*, and *The Himalayan Journal of Contemporary Research*. Also, her poetry has been published in an anthology brought out by Cyberwit publishers, India, in 2016, and *Daath Voyage* journal (ISSN: 2455-7544), 2017. Her poems were shortlisted for the Srinivas Rayaprol Prize (2015).

Puja Sen Majumdar is a research scholar who has recently completed her MPhil from Centre for Studies in Social Sciences, Calcutta. She finished her bachelor's and master's degrees from Department of English, Jadavpur University, Kolkata. Her thesis tried to explore the notion of a deconstructive feminist subjectivity in the works of Donna Haraway. She is interested in literary theory, Marxist and feminist theory, philosophy of literature, and postcolonial studies.

Renée Latchman is an English lecturer at Howard University, Washington, DC. She is from Jamaica, West Indies. She gained her BA degree in linguistics and French from the University of the West Indies, Mona. She completed her MA and PhD in English language and literature at Morgan State University, Baltimore, Maryland, and has special interest in multicultural, Caribbean, and postcolonial literature. Her publications on the Caribbean, women's, and migrant literature include "The Impact of Immigration on Mother-Daughter Relationships and Identity Development in Six Novels of the Caribbean Diaspora" and "The Tradition and Ramifications of Testing in Edwidge Danticat's *Breath, Eyes, Memory*."

Risha Baruah is a research scholar from Department of English, Cotton University. She is engaged in teaching at B. Barooah College affiliated to Gauhati University, and is an academic counsellor of Indira Gandhi National Open University. Her area of interest is ecocriticism, Indian English literature, and Literary Theory. She has authored several scholarly articles and research papers.

Shivani Duggal works as an assistant professor at Jagannath Institute of Management Sciences. She recently completed her MPhil (English) and master's degree (English) from Guru Gobind Singh Indraprastha University (GGSIPU), Delhi. She earned her bachelor's degree (English) from University of Delhi. Her areas of interest are African literature, African diaspora literature, postcolonial studies, and feminist theory. She has been a regular speaker at national and international conferences and has presented papers there.

Shubhanku Kochar works as an assistant professor at University School of Humanities and Social Sciences at GGSIPU, Delhi. He has been teaching since 2012. His areas of interest include African and African diasporic literature along with environment and literature. His works include *Everything Will Be Alright* novel by Author's Ink; *Treatment of Violence: A Reading of Toni Morrison's Selected Fiction* and *An Ecocritical Reading of Alice Walker's Selected Works* by Lambert Academic Publishers of Germany; sixteen research papers in both national and international journals; a chapter for a book on black women writers, which is to be published by Lexington Press; and a play on Indian partition, which is under submission with the publishers. He has also presented various research articles at national and international conferences.

Stephen Ogheneruro Okpadah is a PhD candidate at the Department of the Performing Arts, University of Ilorin, Nigeria. He holds a BA (Hons) degree in theatre arts from Delta State University, Abraka, Nigeria, and a master's degree in performing arts from University of Ilorin, Kwara State, Nigeria. He has published numerous articles in local and international journals such as *Quint Quarterly*, University College of the North in Canada, Routledge Francis, Taylor Group, and *AJPAS: Journal of Pan African Studies, Journal of Media Literacy and Academic Research*, University of SS. Cyril and Methodius in Trnava Slovak Republic, among many others. His areas of research include geopolitics, queer studies, performance aesthetics, and environmental cultural studies.

Suzy Woltmann earned her PhD in literature from the University of California, San Diego, where she teaches literature and writing courses. She specializes in adaptations studies, gender and sexuality, and intertextuality. She has published on topics including gender and sexuality, adaptations, the postcolonial novel, neo-slave narratives, and fairy tales. Her current book project theorizes transformative adaptations that encourage interactive readership.

www.ingramcontent.com/pod-product-compliance
Lightning Source LLC
Chambersburg PA
CBHW020117010526
44115CB00008B/859